CHICAGO PUBLIC LIBRARY
HAROLD WASHINGTON LIBRARY CENTER

R0008258916

Ref. QE 431.2 .W37 1979

Cop.1

Business/Science/Technology
Division

The Chicago Public Library

MAY 8 1981

Received

ROCKS AND MINERALS

Other titles in the *Introducing Geology* series
Editor: J. A. G. Thomas, Verdin Comprehensive School, Winsford, Cheshire.

1 FOSSILS
  F. A. Middlemiss

2 BRITISH STRATIGRAPHY
  F. A. Middlemiss

3 PHYSICAL GEOLOGY
  J. R. L. Allen, FRS

5 FIELDWORK IN GEOLOGY
  I. M. Simpson

INTRODUCING GEOLOGY SERIES, NO. 4

ROCKS AND MINERALS

Janet Watson FRS
Imperial College
University of London

SECOND EDITION

London
GEORGE ALLEN & UNWIN
Boston   Sydney

First published in 1972
Second edition 1979

This book is copyright under the Berne Convention. All rights are reserved. Apart from any fair dealing for the purpose of private study, research, criticism or review, as permitted under the Copyright Act, 1956, no part of this publication may be reproduced, stored in a retrieval system, or transmitted, in any form or by any means, electronic, electrical, chemical, mechanical, optical, photocopying, recording or otherwise, without the prior permission of the copyright owner. Enquiries should be sent to the publishers at the undermentioned address:

GEORGE ALLEN & UNWIN LTD
40 Museum Street, London WC1A 1LU

© Janet Watson, 1979

British Library Cataloguing in Publication Data

Watson, Janet, b. 1923
  Rocks and minerals. – 2nd ed. – (Introducing geology; no. 4).
  1. Mineralogy   2. Rocks
  I. Title   II. Series
  553   QE363.2   78–41295

ISBN 0–04–551031–8

REF
QE
431.2
.W37
1979

Typeset in 10 on 11 point Times by Northampton Phototypesetters Ltd
and printed in Great Britain
by Hazell Watson & Viney Ltd, Aylesbury, Bucks

Contents

Preface	page	7

1 LOOKING AT ROCKS   9
Rocks and their origins   9
Changes in the Earth   10

2 THE THREE CLASSES OF ROCKS   13
Rock-forming processes   13
The three classes of rocks   14
Erosion: the destruction of rocks   16
Earth movements   16
Mobility of the Earth's crust   17

3 MINERALS   18
Minerals in granite   18
Minerals and rocks   18
The composition of rock-forming minerals   19

4 THE SEDIMENTARY ROCKS   21
The source materials   21
Transport and differentiation   21
Environments of deposition   22
Bedding   23
Fossils   24
Changes after deposition, or diagenesis   25
The detrital rocks   26
The chemical–organic rocks   28
Water in sedimentary rocks   31
Petroleum, oil and gas   33

5 VOLCANIC ACTIVITY AND THE IGNEOUS ROCKS   34
Volcanoes and their distribution   35
Volcanic activity   35
Lavas and pyroclastic rocks   38
Igneous intrusions   40
Classification of igneous rocks   41

6 THE METAMORPHIC ROCKS   45
Environments of metamorphism   45
Contact aureoles   46
Dislocation metamorphism   47
Regional metamorphism   47

7 PROPERTIES AND USES OF MINERALS   50
Crystals and crystal symmetry   50
Other diagnostic properties   51
Silicate minerals   53
Non-silicate minerals   54
Ore deposits   56

8 THE USES OF ROCKS AND MINERALS   58
Fuels   58
Building materials   58
Food production   58
Industrial raw materials   58
Ornaments   60

9 THE EARTH AS A WHOLE   61
The crust, mantle and core   61
The continents and oceans   63
The mobile belts   63

INDEX   70

List of Tables

Table		page
1	The composition of the crust	19
2	Common rock-forming minerals	20
3	Environments of deposition of sedimentary rocks	23
4	Classification of volcanic eruptions	37
5	Classification of igneous rocks	41
6	Useful rocks and minerals	59
7	Silicate minerals	65
8	Non-silicate minerals	66
9	Sedimentary rocks	67
10	Igneous rocks	68
11	Metamorphic rocks	69

Preface

Before we can understand the history of the Earth on which we live, we must know something about the materials of which it is made. This book provides a short account of rocks and minerals which are common in the crust of the earth and is intended to meet the needs of candidates studying for the GCE Ordinary Level and similar examinations. Factual information is summarised in a series of tables which are intended to help students to recognise specimens of different rocks and minerals. In the text itself, I have tried to show how the characters of each type of rock are related to the conditions under which it was formed and the figures, which owe much to the skill of Mrs Tamsyn Imison, are designed to illustrate these relationships where possible. I should like to thank Mr J. A. Gee and Miss H. O'Brien of Imperial College who took most of the photographs and to acknowledge permission from the Director, Institute of Geological Sciences, to reproduce seven Crown Copyright Geological Survey photographs.

JANET WATSON
Department of Geology, Imperial College, London.

Preface to Second Edition

In preparing the Second Edition of this book, I have been able to modify the presentation in the light of comments made by some of those who have used it in the classroom. Although the substance remains more or less unchanged, the order in which subjects are tackled has been modified, some gaps have been filled and some irrelevant material has been discarded; the number of illustrations has been increased and most of the diagrams have been redrawn. I should like to thank all those who have offered advice, and especially Mr J. A. G. Thomas, who has given much time and thought to the process of revision.
September 1978

JANET WATSON

Chapter 1

Looking at Rocks

Rocks and their origins

At many places in the hills and around the coasts of Britain, the grass and soil which usually cover the land surface have been worn away to reveal the rocks underneath. Rocks such as these make up the Earth down to a depth of 30 or 40 km. They are of many different kinds and include the raw materials from which we obtain coal, oil, building stone, iron, copper and many other useful substances.

If you live near the coast or in hilly country, you will probably have already looked carefully at rocks exposed near your home. If not, the photo- graphs in this book provide illustrations of some types of rock. Comparison of rocks to be seen in different places soon shows that they are not all alike. Some are soft or crumbly; others are hard and brittle. Some form a regular series of layers, whereas others are uniform or patchy. These variations are not haphazard – the appearance of each rock turns out to have something to do with the way in which that rock was formed.

A first example of this fact is provided by some rocks exposed in the cliffs near North Berwick (Fig. 1.1). These rocks form regular layers which are

Figure 1.1 The ripple marks on this rock surface from North Berwick show that the rock was formed originally from sand deposited under water. The rock has been tilted since its formation. (Crown copyright)

Figure 1.2 A volcanic bomb from the Canary Islands.

tilted at an angle of about 60° towards the observer. The surface of the nearest layer shows a wavy pattern of low ridges and troughs spaced at intervals of 10–20 cm. This pattern may remind you of the rippled surface of a sandy beach seen at low tide – indeed, the shape and spacing of the ridges on the rock surface could be matched exactly on many beaches. We know that sand ripples are formed by the motion of shallow water, which heaps loose sand grains into a regular arrangement of ridges as gentle currents wash to and fro. So closely do the ripple marks on the rock surface shown in Figure 1.1 resemble the sand ripples on a beach that we can be reasonably sure that they were developed in the same way. If this conclusion is correct we must assume that the ripple-marked rock was originally formed as a layer of loose sand grains lying in shallow water. A close look with a lens at a broken surface of the rock would show that it is, indeed, made largely of sand grains, but these grains have been firmly welded together to make the rock hard and cohesive. Furthermore, we can be certain that the ripple-marked layer must, like the surface of a beach, have been almost horizontal at the time when it was formed, but this layer is now steeply tilted. If we accept the idea that the rock layer was formed from sand on a sea floor, it follows that this layer has been hardened and tilted at some time after the ripple marks were formed.

Figure 1.2 shows a rock of a very different kind – a compact, elongated lump with a crust on the outside cracked like the crust of a cottage loaf. Rocks of this kind can be picked up on the slopes of active volcanoes such as those of the Canary Islands, and if you have watched films of volcanic eruptions you will probably be able to guess how they were formed. From time to time the hot gases inside the volcano build up sufficient pressure to force their way to the surface. The explosive outrush of gas carries up with it both shattered rock fragments and white-hot blobs of frothy molten lava. These blobs are streamlined or twisted in flight and the gas bubbles they contain are frozen in as the molten material cools on contact with the air. Rocks formed from the blobs of molten lava thrown up during an explosive volcanic eruption are called volcanic bombs. The example illustrated in Figure 1.2 shows clearly the shape of the solidified lump and a part of its twisted outer crust, cracked by the expansion of gases which were trapped in the still-fluid interior.

These examples show that something can be learned about the origins of rocks by looking not only at the material of which they are made but also at the way in which it is arranged. Geologists – the scientists whose business it is to study the Earth – use evidence of this kind to build up ideas on how the rocks which make the Earth have been formed, and how the Earth as a whole has changed since the time of its origin. This book is intended to provide a description of the most common rocks and an account of the rock-forming processes responsible for their origin.

Changes in the Earth

A few kilometres east of the city of Nairn, in north-east Scotland, lies a desolate area of sand dunes, now largely covered by rough grass and conifer plantations. In the Middle Ages this area, the Culbin Sands, was a fertile region containing several farms. Sand dunes then occupied a belt of land nearer the sea, well to the north. Year after year, strong winds blowing from the sea carried loose sand with them, so that the dune belt gradually advanced over the farms. A succession of heavy storms in the 1690s led to the burial of fields and buildings, and by the end of the 17th century most of the farms had been abandoned. In recent times the advance of the dunes has been halted by the planting of grass and trees whose roots bind and stabilise the sand, but the old houses and farm buildings are still buried.

The story of the Culbin Sands illustrates the fact that the Earth is always changing. You can convince yourself of this fact if you look carefully

CHANGES IN THE EARTH 11

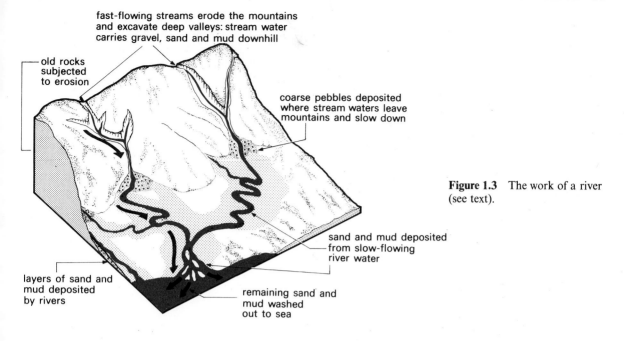

Figure 1.3 The work of a river (see text).

at any country landscape you know well. After a few days of heavy rain, for example, river water usually becomes brown and muddy. The rainwater trickling downhill has picked up dust and soil which is washed into tributary streams and thence into the main rivers. Much of this mud settles out later in quiet waters near the river mouth (Fig. 1.3).

The effect of these everyday events is to remove particles of sand and dust from the hilly areas upstream and to deposit these same particles as layers of sand or mud many kilometres downstream (Fig. 1.4). These processes seem unspectacular because they operate slowly, but because they are going on continuously, century after century, they are capable of bringing about enormous changes. The Mississippi River in the USA, one of the world's largest rivers, carries two million tonnes of sand and mud towards the sea each day and has dumped a pile of sediment at its mouth to form the Mississippi delta which juts out into the Gulf of Mexico.

Changes of an entirely different sort result from earthquakes. The violent shaking produced by an earthquake is due to the sudden movement of rocks beneath the surface along a fracture which is called a fault. These earthquakes show that the hot inner part of the Earth is under stress. The disturbances they produce are recorded by fracturing or tilting

Figure 1.4 Mud deposited in the estuary of a river in Northern Ireland. The whelk and cockle shells will become fossils when they have been buried by new layers of mud. The cracks are the result of shrinkage during periods when the mud is exposed to the air. (Crown copyright)

of rock layers and often also by sagging or bulging of the land surface above the earthquake focus.

These signs of activity in the Earth show that it is always changing. The results seldom seem dramatic to us, because a human lifetime is very short compared with the length of **geological time** – that is, with the period of four to five thousand million years since the Earth originated. If we go back even to the earliest days of the human race, we have covered only a few million years representing the very last stage of the history of the Earth. There are of course no written records of the Earth's history, but the rocks themselves provide clues to this history. To understand them, we must consider in more detail the ways in which they have been formed.

Because man, like all other animals, lives at the surface of the Earth, it is natural that geologists should know far more about the outer parts of the Earth than they do about its interior. In most of this book, we shall therefore be dealing with rocks that form the outermost solid crust of the Earth. We can think of this **crust** as a rocky layer, 30–40 km thick under the continents and 5–10 km thick under the oceans. Beneath it are rocks of a different kind, which are seldom seen at the surface. This inner shell of the earth, which is known as the **mantle**, encloses a central **core** made largely of iron and nickel. Most of what we know about the mantle and core comes not from direct observation but from instruments designed to investigate the properties of rocks which lie far beneath the surface. We will return to this subject in the last chapter of the book.

Chapter 2

The Three Classes of Rocks

Rock-forming processes

All the solid materials of the Earth's crust, except those made directly by man and those forming the bodies of animals and plants, are classed by geologists as **rocks**. In ordinary life we think of a rock as something hard, but geologists extend the term to cover sands, clays and other soft or unconsolidated natural substances. Many of these loose materials are in fact hard rocks in the making, as we shall see later.

In Chapter 1, we have seen that rivers carry towards the sea mud and sand picked up in their higher reaches, and that some of this material settles out again in the river bed or in the sea near its mouth (Fig. 1.3). We have seen, too, that the wind can pick up sand and dust from one place and redeposit it somewhere else. The river muds and the dune sands are rocks in the making. The river and wind, respectively, are the agents responsible for their formation and in order to understand the processes by which mud and sand were deposited, we should have to study the river and wind systems.

A simple experiment will illustrate some of the factors involved in deposition from water. Half fill a glass vessel (a jam-jar will do) with water and put into it a handful of crumbly soil. Shake it gently to get rid of trapped air bubbles. You will find, as a rule, that most of the large particles fall almost immediately to the bottom under the influence of gravity, but that the water turns cloudy because the finest particles of mud remain in suspension. If the jar is left untouched for several hours, the mud finally settles out to form a dark layer on top of the sediment (Fig. 2.1). Even a small movement of the jar will cause some of this layer to swirl up into the water again. A much stronger shake is needed to disturb the sand underneath. If you think about this experiment, you will realise that it can tell you two things about the behaviour of sediment in water:

(1) The coarser grains settle out more quickly than the fine mud particles. This is because

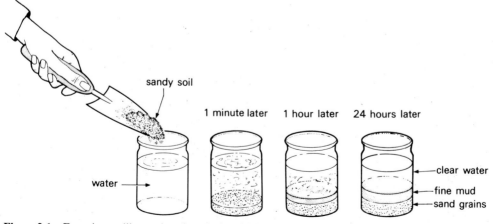

Figure 2.1 Experiment illustrating the way in which sand and mud settle out from water.

Figure 2.2 Layers of sand deposited one above the other by a stream crossing a beach. The lower layers are exposed where sand has fallen away from a steep slope. (Crown copyright)

the larger grains fall under their own weight and quite a strong movement of water is needed to keep them in suspension; on the other hand, minute mud flakes are so light that they remain suspended for a long time even in still water. If we apply this observation to a river, we can guess that as the current slows down (say where the river enters a lake), the coarser sand which it carries will settle out quickly, whereas the mud may be carried further downstream; thus, the grain size of an old river sediment will tell us something about the rate of flow of the river water that deposited it.

(2) When particles of any kind are deposited from water, the grains that reach the bottom first lie at the base of the pile and those that settle out last lie at the top. The sand or mud deposited on a river bed often shows a roughly horizontal layering, due to slight variations in grain size or in colour (Fig. 2.2). From the results of the jam-jar experiment, we can say that the higher layers were deposited after the layers below. Differences of grain size between successive layers suggest that the layers were deposited during periods when the river flowed at different rates.

The three classes of rocks

Observations and arguments like those outlined above are used by geologists to establish the links between the nature of the rocks they can study and the operation of the appropriate **rock-forming process**. Because rocks can be formed by many different processes, our first need is to group them according to their characteristic features and their modes of origin. A threefold grouping into sedimentary, igneous and metamorphic rocks provides the best starting point.

Sedimentary rocks
These rocks are formed at the surface of the Earth by the accumulation of particles of solid matter and by the precipitation of substances dissolved in water. Water, ice and air – the rivers, seas, glaciers and wind – are responsible for transporting pebbles, sand and silt over the surface and finally for sweeping them into hollows on the land or into the depths of the sea. The formation of sedimentary rocks is thus bound up with happenings familiar to everyone – a heavy shower washing soil into a stream sets off a train of events which may end with the settling-out of mud in a pool downstream and the formation of a new rock.

Sedimentary processes and sedimentary rocks are described in Chapter 4.

Igneous rocks
These are formed by the solidification of very hot molten rock material or **magma** ('igniis' is the Latin word for 'by fire'). Such magma is formed where high temperatures are reached near the base of the crust or in the mantle and it is squeezed up towards the surface until it solidifies on contact with the cold crust or with the sea water or air. Molten magma feeds volcanoes and is responsible for the phenomenon of volcanicity or volcanic activity. Portions of the magma which escape at the Earth's surface solidify to form **lava**. Those which do not reach the surface consolidate in spaces within the crust to produce **intrusive igneous rocks**. Volcanicity and the igneous rocks are described in Chapter 5.

Metamorphic rocks
These are rocks formed from sedimentary or igneous parents which have been modified by high temperatures and pressures inside the crust. The process of **metamorphism** (the term is derived from Greek words meaning 'change of form') does not lead to wholesale melting. Metamorphic rocks therefore usually retain some of the features of their sedimentary or igneous parents. But the materials of which they are made have recrystallised to form new compounds in equilibrium at high temperatures and pressures (see Ch. 6).

The three classes of rocks are formed in *different geological environments*. Sedimentary rocks are deposited at the Earth's surface by processes operating at atmospheric temperatures and pressures. Igneous rocks are formed at temperatures of 800–1200°C and the molten magma from which they are produced is generated deep in the Earth, usually more than 30 km below the surface. Metamorphic rocks, also the products of heat, are formed at temperatures below the melting points of most rocks and at depths of 5–45 km. A diagrammatic section of the outer part of the Earth (Fig. 2.3) shows the environments in which rocks of the three classes are formed.

As you look at Figure 2.3, you will probably wonder how it is that any rocks formed inside the Earth – the intrusive igneous rocks and the metamorphic rocks – come to be exposed at the surface where they can be examined by geologists. The answer to this question arises from two facts which were mentioned in Chapter 1. First, there are destructive forces at work at the Earth's surface

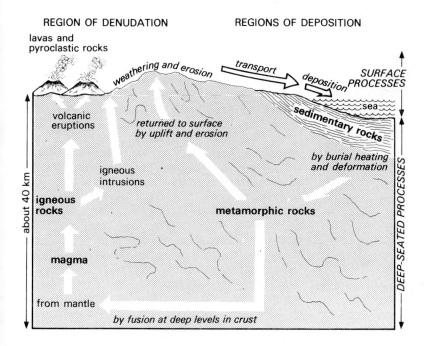

Figure 2.3 The three classes of rocks.

which wear away the outermost parts of the lands to expose the rocks underneath. Secondly, the Earth is subject to internal stresses which break and crumple its rocks and which can upheave whole regions to form new lands and new mountain chains. We will take these two points in turn.

Erosion: the destruction of rocks

Stream waters in flood, as we have seen, sweep away sand, mud and even gravel and boulders. Armed with these lumps and grains of solid matter, fast-flowing water has great power. The bigger fragments, battering at the floor and banks of a stream, break off new fragments which in turn are caught up by the water. In this way, after many years, the stream wears down its course and begins to run in a deep gorge. After a while, undercutting by the stream, aided by fracturing and slipping of its banks, widens the gorge into a steep-sided valley while tributary streams cut new trenches in the valley slopes. In this way, the old hills are gradually worn away and rocks that originally lay far below the surface are exposed (Fig. 2.4). The process of wearing away of the surface, by rivers or by other agents, is called **erosion**. If you look again at Figure 1.3, you will realise that the sand and mud that are deposited near the mouth of the river represent the products of erosion near its source. The river acts as an **agent of erosion** in mountain regions, where it is flowing rapidly, and as an **agent of deposition** near its mouth where its waters have slowed down.

Many other natural processes have the effect of wearing away the rocks of the lands. In mountain regions, the freezing of water in cracks causes expansion which gradually shatters the rocks (the same thing can happen in newly made concrete, which is why the concrete foundations of a building cannot be laid in frosty weather). Loose rock fragments and soft soil or mud may slide downhill under their own weight when saturated with rainwater. The waves of the sea undermine cliffs and gradually eat their way into the land.

The general effects of these processes are to wear away mountains, to smooth down the contours of the land. If erosion were the only process at work, we might expect that all the continents would have been reduced by now to low, featureless plains. But in fact, mountains several kilometres in height rise from the continents.

Earth movements

The tallest mountains on Earth are the Himalayas. On the flanks of Mount Everest, the highest of the Himalayan peaks, sedimentary rocks have been found which contain fossilised shells very similar to those of shellfish that can be found in the sea today. The occurrence of these fossil shells demonstrates that at least some of the rocks that make Mount Everest and the other Himalayan peaks were formed in the sea. Now they are 7 km above sea level – how did they reach this height? The answer seems to be that the whole Himalayan range has been forced bodily upward as two parts of the Earth's crust were squeezed together. The upheaval of this huge mountain mass was followed by rapid and deep erosion, as streams carved new valleys in the steep slopes. The layers of fossiliferous sedimentary rocks were thus exposed in the rocky walls of the newly formed valleys.

The **earth movements** that led to the upheaval of the Himalayas were of the kind associated with repeated earthquakes (p. 11) and even today the Himalayan region is subject to earthquakes. Other regions have been subject to similar disturbances in earlier geological times. In Britain the forces responsible for such movements have long since died away, but their effects can still be recognised by geologists, because rocks originally formed far

Figure 2.4 Erosion by streams. The slopes of this mountain in Iran show parallel gullies eroded by small streams; the cliff in the foreground is the wall of a river-gorge which cuts through the mountain. (Aerofilms Ltd.)

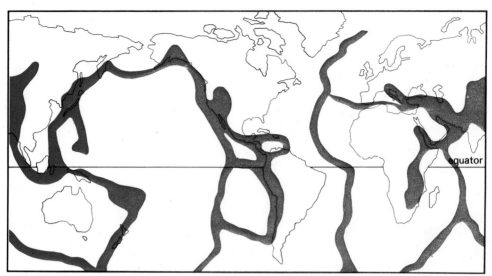

Figure 2.5 Earthquake belts. The coloured areas mark the parts of the Earth most often affected by earthquakes.

below the surface of the Earth are now exposed at the surface. The rocky cliffs and tors of Land's End in Cornwall and Dartmoor in Devon, for example, are made of granite which was originally intruded several kilometres beneath the land surface. Gradual upheaval of the Earth's crust in these regions, many millions of years ago, was followed by erosion by rivers, wind and sea which stripped off the overlying rocks. Earth movements of this kind complete the **cycle of rock-forming processes** illustrated in Figure 2.3.

Mobility of the Earth's crust

The most dramatic kinds of earth movements are the sudden jolts that cause earthquakes (p. 11). A world map marking the sites of large earthquakes shows that some parts of the Earth are much more active than others (Fig. 2.5). Earthquakes are rare in Britain and those recorded in historic times have seldom done anything worse than shake down a few chimney-pots. This illustrates the fact that the British area is **geologically stable** – that is, it is not subject to strong earth movements. On the other hand, large and very destructive earthquakes have been recorded again and again in the Mediterranean region and along the Pacific coasts of North and South America. These regions are **geologically active**, and because powerful earth movements take place in them, the crust in these parts of the Earth is said to be **mobile**. The map in Figure 9.3 (p. 64) shows the distribution of **mobile belts** in the Earth. Some of these lie beneath the oceans; others form rugged mountainous country, including the Himalayan Mountains. As was mentioned above these mountains have been forced bodily upward as they were compressed between adjacent regions. These mountainous mobile belts are therefore called orogenic (mountain-building) belts. We will return to them in Chapter 9.

Chapter 3

Minerals

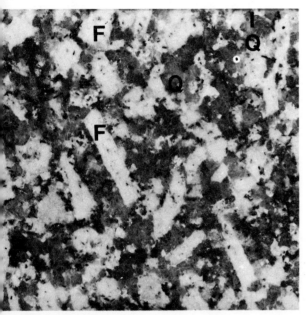

Figure 3.1 Granite: a coarse-grained igneous rock containing the minerals quartz (Q), feldspar (F) and mica. The large feldspars are 1–2 cm long.

Minerals in granite

The intrusive igneous rock **granite**, which was mentioned near the end of Chapter 2, is often quarried for use as an ornamental stone. Slabs of granite were used to face many large public buildings erected in the 19th and early 20th centuries, and form the headstones of graves in many churchyards. Seen from a distance, granite is a light grey, pink or buff rock, with a slightly mottled surface. A close look at a specimen, using a hand lens for magnification if necessary, will show that the mottling results from the presence of grains of several different kinds. These grains represent the different **minerals** found in granite (Fig. 3.1).

The most common mineral is opaque and white, buff or pink in colour; it tends to break along parallel cracks, giving smooth shiny surfaces. This mineral is **feldspar**. Associated with it are shapeless grains of a transparent, glassy-looking mineral known as **quartz**, and black or silver flakes of the mineral **mica**. When a chip of granite is gently crushed, the grains fall apart and the three minerals can be separated from one another. Using a lens, you may be able to see that the quartz grains have an irregular shape, whereas many of the feldspars form tiny brick-shaped particles, because of their tendency to break along smooth surfaces. Micas are almost always flaky and if you can separate a large flake you will find that it bends easily, but tends to spring back into shape when released. Mica can be scratched by a knife; feldspar and quartz cannot.

Minerals and rocks

Minerals are the natural units that make up most rocks. Each mineral has a more or less constant chemical composition which can be expressed by a chemical formula. It also has a definite atomic structure – that is to say, the various atoms making up the mineral substance are arranged in a regular pattern which is the same for all samples of a single mineral species. The geometrical regularity of the atomic lattice is reflected in the outward shapes assumed by minerals. When free to develop without interference, many minerals form **crystals** bounded by symmetrically arranged plane surfaces (Fig. 3.2). Under natural conditions, the ideal crystal forms are seldom achieved and most minerals are therefore seen as irregular grains or as grains only partly bounded by crystal faces.

Most rocks contain several different minerals, and the nature of the **rock-forming minerals** often helps to define a particular type of rock. Granite, for example, can be defined as an intrusive igneous

THE COMPOSITION OF ROCK-FORMING MINERALS

Figure 3.2 Crystals: a group of quartz crystals. (Crown copyright)

rock made largely of quartz, feldspar and mica (Fig. 3.1). Before we can give a complete description of a rock, we must therefore be able to recognise the minerals of which it is made and to know something about their composition. Some minerals, moreover, are interesting and important in their own right. Diamond, sapphire, ruby and other gemstones, for example, are naturally formed minerals which are valued for their colour or for the fact that they can be elaborately cut and polished. Diamond also has many commercial uses because of its exceptional hardness, and other minerals are valued as the sources from which metals such as chromium, tin or nickel are extracted.

For all these reasons, geologists need to study minerals as well as rocks. The last section of this chapter is intended to introduce the names of the rock-forming minerals that will be mentioned in the next few chapters, and to provide some background information about their chemical compositions. Further details are given in Chapter 7.

The composition of rock-forming minerals

The minerals that can be found in rocks like granite are made up of elements which are common in the outer parts of the Earth. About 99% of the rocks in the crust are in fact made up of only eight elements, as is shown in Table 1, below.

Table 1 The composition of the crust.

Eight **major elements** form about 99% of the crust. They are, in order of abundance:

Element	% (by weight)
oxygen (O)	47
silicon (Si)	28
aluminium (Al)	8
iron (Fe)	5
calcium (Ca)	3.5
sodium (Na)	2.8
potassium (K)	2.6
magnesium (Mg)	2

All other elements put together form only about 1% of the crust. The most important of these **trace elements** are titanium (Ti), hydrogen (H), phosphorus (P), manganese (Mn), sulphur (S) and carbon (C).

MINERALS

The two most abundant elements in the outer parts of the Earth are **silicon** and **oxygen**. These elements easily form chemical compounds with each other and with aluminium, iron and other common elements, so it is natural to find that most rocks are made of silicates or oxides of aluminium, iron, calcium, sodium, potassium or magnesium.

In Table 2, the common rock-forming minerals are named and grouped according to their chemical composition. By looking back to this table, you can learn something about the compositions of the rocks mentioned in the next few chapters. For example, the rock named granite discussed earlier in this chapter consists mainly of quartz, feldspar and mica; it must therefore contain a high percentage of silicon, oxygen and aluminium, since they are the principal elements present in these minerals. You will find more details in Tables 7 and 8 on pages 65 and 66, and a general account of the properties of common minerals in Chapter 7.

Table 2 Common rock-forming minerals.

Mineral	Chemical classification	Principal elements	Common in:
Silicates			
quartz	silica, SiO_2	Si, O	igneous, sedimentary and metamorphic rocks
feldspar	silicate	Si, O, Al, Na, Ca, K	igneous, sedimentary and metamorphic rocks
mica	silicate	Si, O, Al, K, OH (Mg, Fe)	igneous, sedimentary and metamorphic rocks
hornblende (amphibole family)	silicate	Si, O, Al, Fe, Mg, Ca, OH	igneous and metamorphic rocks
augite (pyroxene family)	silicate	Si, O, Fe, Mg, Ca, (Al)	igneous and metamorphic rocks
olivine	silicate	Si, O, Fe, Mg	igneous rocks
garnet	silicate	Si, O, Al, Fe, Mg, Ca	metamorphic rocks
clay minerals	silicates	Si, O, Al, OH	sedimentary rocks
Non-silicates			
calcite	carbonate	Ca, C, O	sedimentary and metamorphic rocks
aragonite	carbonate	Ca, C, O	sedimentary rocks
dolomite	carbonate	Ca, Mg, C, O	sedimentary and metamorphic rocks
magnetite	oxide	Fe, O	sedimentary, igneous and metamorphic rocks
haematite	oxide	Fe, O	sedimentary, igneous and metamorphic rocks
halite (rock salt)	chloride	Na, Cl	sedimentary rocks
fluorspar	fluoride	Ca, F	veins in sedimentary or igneous rocks
gypsum	sulphate	Ca, S, O, OH	sedimentary rocks
barytes	sulphate	Ba, S, O	veins in sedimentary rocks
pyrite	sulphide	Fe, S	sedimentary, igneous and metamorphic rocks

Note: in Column 3, OH (the hydroxyl radical) denotes oxygen in combination with hydrogen.

Chapter 4
The Sedimentary Rocks

All the rock materials that appear at the land surface are exposed to the effects of rain, wind, frost and sunshine – in fact, they are exposed to the weather and after a time they become weathered. Old minerals decay and are replaced by new ones, minute cracks are opened, fractures are widened and deepened so that, in the end, weathered rocks disintegrate into loose material that is easily washed away or blown away. As was noted on page 14, the processes of wearing away (**denudation** or **erosion**) of the land surface, which are begun in this way, are continued by the agencies of rivers, winds, glaciers and seas. All these **agents of erosion** are controlled in the end by the force of gravity. Rivers flow out of the highlands towards the sea, carrying with them the debris resulting from erosion and the dissolved salts formed by weathering. Glaciers and ice-sheets move slowly seaward until melting sets in. The wind is less strictly bound by gravity, but the sand and dust it carries nevertheless settle out in sheltered spots under the effect of gravity. As a result of all these processes, the mantle of loose debris produced by weathering and erosion is transported piecemeal from highlands to lowlands and from lowlands to the sea. The agents of transport are essentially the same as the agents of erosion – water, ice and wind, aided by gravity.

It follows, then, that the lands are generally **regions of erosion** from which rock material is lost. Coastal plains, deltas and shallow seas are the sites to which eroded material ultimately moves; they are **regions of deposition** and it is there that most sedimentary rocks are made (Fig. 2.3).

The source materials

Although weathering and erosion of older rocks provide the bulk of the material incorporated in sedimentary rocks, they are not the only contributors. Volcanic eruptions expel enormous amounts of ash which may settle where it falls or be carried to areas of deposition by wind or rivers. Animals and plants contribute more material than might be imagined. The plants of forests and swamps decay almost *in situ* to give peat, coal or oil. Shellfish, corals and other marine invertebrates extract calcium carbonate and other substances from sea water to build shells and skeletons which may be buried as fossils in the accumulating sediments.

Transport and differentiation

The raw materials from which the sediments are to be made sometimes travel hundreds of kilometres from their source to their final destination. At every stage of their journey, they are roughly sorted out according to their size, shape, density, composition and solubility. To take one example, the load of rock material carried by a river is sorted into four fractions:

(1) **boulders and large grains** (diameter greater than 2 mm) too heavy to be lifted by the water are rolled along the river bed;
(2) **lighter particles** of sand ($2-\frac{1}{16}$ mm) travel downstream with a skipping motion in a series of short hops;
(3) **the finest particles** are carried in suspension and only settle out in calm water; and
(4) **dissolved material** such as calcium bicarbonate and iron salts may remain in solution indefinitely.

The effect of sorting (such as that which goes on in rivers) is to reassemble the materials supplied by the source areas into fractions differing from each other in composition, grain size and mode of occurrence. Figure 4.1 illustrates the effects of **differentiation** of the raw materials and introduces

22 THE SEDIMENTARY ROCKS

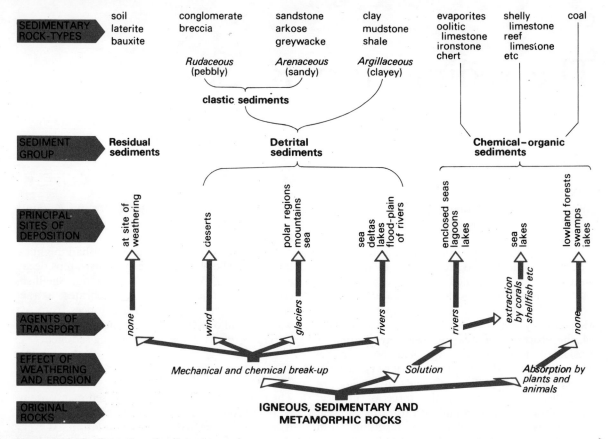

Figure 4.1 The formation of sedimentary rocks.

some new terms for the sediments that result from the process. Three main divisions may be mentioned at this stage: the **detrital sediments,** the **chemical–organic sediments** and the **residual sediments.** The detrital sediments are formed by the accumulation of transported particles of old minerals and of new minerals produced during weathering. The chemical–organic sediments are formed from materials transported as colloids or in solution, and from materials manufactured by plants and animals. These two divisions together form the vast majority of sedimentary accumulations. The residual sediments are materials left behind at the site of weathering, either because they were difficult to transport or because they were stabilised by a plant cover. **Soils,** in which organic material and bacteria are mixed with the products of weathering, are the most important substances in this category.

These major groups can be sub-divided to give a classification of all sedimentary rocks as shown in Figure 4.1. The top lines of this Figure show the rock groups that will be described in the next few pages. Table 9, page 67, summarises details about these rocks.

Environments of deposition

You will realise at once that the **agents of deposition** are essentially the same as the agents of transport. A river flowing into a lake deposits sand and mud as its current slackens. A melting glacier dumps a jumble of boulders, sand and mud on the valley floor. The wind drops, and sand and dust settle out from it. Many such incidents turn out to be only temporary halts: a storm may stir up mud from the lake floor, streams of meltwater may carry away the glacial debris, the wind may rise again, and so on. Sooner or later, most sedimentary material makes its way to the lowlands or the sea, by far the most important environments of deposi-

Table 3 Environments of deposition of sedimentary rocks.

ENVIRONMENT OF DEPOSITION			COMMON SEDIMENTARY ROCKS
Sea			
shallow seas (continental shelf)	Littoral or shoreline (beaches, sandbanks, tidal flats)		conglomerate, sandstone, shale (evaporites)
	neritic or shallow waters	open seas	quartzite, current-bedded sandstone, shale, organic and chemical limestones
		enclosed seas	black shale, source rocks of bitumens and oils, evaporites
deep seas	in mobile belts		greywacke and other turbidites, shale
	in stable areas		shale, deep-water limestone, sometimes greywacke
abyssal seas			calcareous ooze, siliceous ooze, red clay
Mixed environments			
deltas			mainly sandstone, shale, coal
estuaries, lagoons			shale
Land			
floodplains of rivers			conglomerate, sandstone, shale
lakes	with outlet to sea		sandstone, shale, freshwater limestone
	in basins of interior drainage		sandstone, shale, evaporites
deserts			sandstone, conglomerate, breccia
piedmont or mountain foot			conglomerate, breccia, arkose, sandstone
areas of glaciation			tillite

tion. Some material, however, remains in valleys, lakes, deserts and swamps, giving sediments of distinctive kinds.

We can therefore group the sediments *in relation to the environments of deposition* and the principal agents of deposition as shown in Table 3. The new terms introduced in this Table will be explained in the descriptive sections on pages 26–31.

Bedding

Sediments are deposited from water, ice or air on the land surface or the sea floor. They are built up layer by layer, each new deposit or **bed** resting on top of those previously laid down. The layered structure or **bedding** that results from this mode of accumulation is characteristic of the class of sedimentary rocks (Figs 2.2 and 4.2). The **bedding planes** are the surfaces separating successive beds in a pile of sediment: each was for a short time the surface of the Earth, and remains in the sedimentary rocks as an easily recognisable surface (often separating materials of slightly different colour or composition) which marks a break or change in the supply of sediment.

Figure 4.2 Bedding in sedimentary rocks, Pembrokeshire. (Crown copyright)

24 THE SEDIMENTARY ROCKS

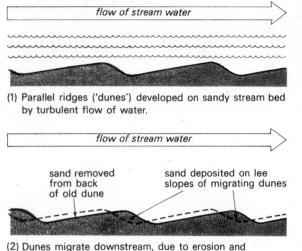

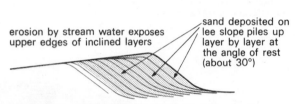

Figure 4.3 Cross bedding: diagrams illustrating the effects of currents on deposition.

From the nature of the process of deposition, it follows that most beds are laid down as almost horizontal sheets whose lateral extent is very much greater than their thickness. Variations on this common form reflect the conditions of deposition. Sediment deposited in a creek or stream, for example, may form a ribbon-like body no broader than the stream and lenticular in cross section. Sediment deposited on a sloping surface forms a bed which is inclined to the horizontal. Inclined bedding is seen, for example, in sand dunes, where the bedding surfaces are almost parallel to the lee slope of the dune, or in many shallow-water sand deposits; it is usually called **cross bedding, current bedding** or **false bedding**.

Figure 4.4 Cross bedding in a water-laid sandstone (cf. Fig. 4.3). (Crown copyright)

The formation of cross bedding depends on the action of eddies as water or wind moves over a surface on which loose grains are lying, as explained in Figure 4.3. Cross bedding in a water-laid sandstone is illustrated in Figure 4.4. You will see from this photograph that each set of inclined beds is cut across by the base of the overlying bed. This arrangement makes it possible to recognise the original top of a series of cross-bedded sedimentary rocks when these rocks have been tilted or folded (see p. 46).

Fossils

The shells and skeletons of animals and the seeds, stems and leaves of plants are often mixed with sand and mud, and become incorporated in the sedimentary pile as **fossils**. Some chemical–organic sediments (p. 26) are made almost entirely of material derived from plants and animals (Fig. 4.8). **Shell sands** made of broken shell fragments provide one example. **Coral reefs** are built entirely by the activity of colonial corals and other plants and animals that are able to precipitate calcium carbonate from sea water. Certain animals and plants are thus rock-forming agents. Many others

Figure 4.5 Trace fossils. The vertical tubes in the sandstone bed below the hammer represent the burrows of worms probably living between tide marks. (Crown copyright)

simply get caught up in the process of deposition and are entombed accidentally. Whichever category they belong to, fossils are important to the geologist in two ways.

First, the nature of the fossils in a sediment *throws light on the environment of deposition*. Remains of marine animals such as sea urchins show that the sediments containing them were laid down in the sea. Reef limestones formed by colonial corals indicate accumulation in the shallow, clean and warm seas in which corals flourish. Infilled burrows formed by worms similar to the lugworm (which lives between tide marks) suggest deposition in the intertidal zone of a beach or estuary. Structures such as the burrows illustrated in Figure 4.5, which are formed by the activities of animals, are called **trace fossils**. Although evidence of this kind can be very useful, it must be looked at with care because animals and plants are not always fossilised in the places where they actually lived. For example, dead animals and plants may be swept away by rivers in flood and deposited in lakes or shallow seas: shell banks may be piled up by currents far from the place where the shellfish lived – in other words, the evidence of fossils must be looked at in the light of all the information available about the history of deposition.

The second value of fossils in the study of sediments is as *indicators of geological age*. Since life began, organisms have undergone progressive changes in structure adapting them to their conditions of life. As a result of this process of **organic evolution**, fossils of different ages record different stages of evolution. The subject of **stratigraphy** – the study of sequences of sedimentary rocks deposited over long periods of time – has grown up largely because it is possible to identify rocks of different ages by means of the fossils they contain. This is a whole subject in itself and is dealt with in *British Stratigraphy* by F. A. Middlemiss (Introducing Geology series, no. 2).

Changes after deposition, or diagenesis

Most of the sediments that accumulate in the seas, lakes and rivers are soft, unconsolidated and full of water. As material piles up on top, they are compressed and compacted and lose much of their water. The remaining spaces between grains – the pore spaces – are often filled in by precipitation of a cement of calcium carbonate, iron oxides or silica. The combined results of burial, dewatering and cementation are to convert the soft, uncon-

Figure 4.6 A conglomerate from the Old Red Sandstone of the Shetland Islands. The pebbles in this conglomerate are only partly rounded. (Crown copyright)

solidated sediment into a denser, harder and more coherent sedimentary rock. The ancient deposits of early geological times studied by stratigraphers have undergone these changes which collectively constitute the process of **diagenesis**.

The detrital rocks

Detrital rocks (Figs 4.4 and 4.6) are formed by the accumulation of solid particles derived from the breakdown of pre-existing rocks during weathering and erosion. Many of the coarser particles – boulders, pebbles and sand-grains – are rock fragments or mineral grains which have suffered very little chemical change during erosion and transport. They represent either stable materials such as quartz, which do not react easily, or materials which have not travelled far from their source (see Fig. 4.6). Where mineral fragments derived directly from older material predominate, the rocks are called **clastic** sediments. On the other hand, most of the finer mud consists of new minerals formed by chemical reactions during weathering. Many minerals of igneous or metamorphic origin originally formed at high temperatures are easily broken down on exposure to surface waters, releasing some elements in solution and forming minerals stable under atmospheric conditions. The chief of the new minerals produced by chemical weathering are the **clay minerals**, silicates containing aluminium and hydroxyl (OH), whose sub-microscopic flakes are the principal ingredients of clays.

Rudaceous (pebbly) rocks

The coarsest grades of clastic rocks contain many grains, pebbles and boulders more than 2 mm in diameter. The size and weight of the particles are such that only powerful agents of transport can carry them and they often accumulate close to the source. **Breccias** consisting of angular rock fragments most often represent **screes**, the deposits that pile up under gravity at the base of cliffs and on steep valley sides. **Conglomerates** (Fig. 4.6) consist of rounded fragments, smoothed by constant collisions during transport. They are common deposits of fast mountain rivers and of beaches where the battering of the waves effects a high degree of rounding.

Boulder clay (till) is the deposit of ice sheets or glaciers and is distinguished by the fact that the components are almost unsorted; moving ice can carry fragments of all sizes which are dumped indiscriminately on melting, and boulder clay therefore shows a medley of boulders – often planed off by abrasion at the ice-foot – scattered through a sandy or clayey matrix. Most boulder clay in Britain was deposited in the geologically recent Pleistocene ice age, when ice sheets covered much of the land; it is still soft and unconsolidated. Similar deposits formed during earlier ice ages are hard and coherent and are known as **tillites**.

Arenaceous (sandy) rocks
Detrital sediments of sand grade ($\frac{1}{16}$ – 2 mm) may be deposited by rivers, seas and wind (Fig. 4.4). The most abundant component of these sandy or arenaceous rocks is quartz, the most stable of common rock-forming minerals. Many **sandstones** consist almost entirely of quartz; when cemented by silica, these rocks are termed **quartzite**. Small quantities of minerals such as feldspar and mica are present in all sandstones, and feldspars form a respectable proportion of the rocks known as **arkoses** and **greywackes**. Arkoses are the products of rapid erosion which left no time for chemical weathering. Greywackes are badly sorted sediments with a wide range of grain size and a good many fragments of feldspar and other unstable minerals or rocks.

Water-laid sandstones are deposited in the channels of rivers, on beaches and in shallow seas where the motion of currents and waves allows sand-sized particles to accumulate, but keeps finer silt and mud in motion. Cross bedding (Figs 4.3 and 4.4) or ripple-marks (Fig. 1.1) are characteristic of many shallow-water sandstones. Fossils in these rocks tend to be broken and abraded.

Greywackes often show **graded bedding** in which coarser grains are concentrated at the base of each bed and finer grains towards the top. Each graded bed records the rapid arrival in the basin of deposition of a lot of poorly sorted sediment which settled under gravity, with the larger grains near the base, in the manner illustrated in Figure 2.1; storms which stir up loose sediment already on the sea floor may initiate the formation of such a graded bed. A series of beds of graded greywacke is often piled up near the base of a submarine slope down which sand and mud originally deposited nearer the land has cascaded as a 'slurry' of water, mud and sand when disturbed by storms or earthquakes. Such underwater cascades of dense muddy water are called **turbidity currents** and the sediments produced by them are known as **turbidites** (Fig. 4.7). The upward reduction of grain size in a graded bed of greywacke makes it possible to recognise the original top of a succession of greywackes which has been tilted or folded after deposition (see p. 46).

Wind-blown or **aeolian sandstones** are formed mainly in desert regions where the lack of plant cover exposes weathered rock and soil to erosion

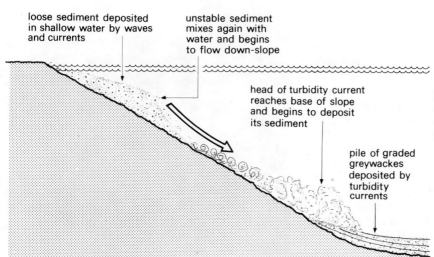

Figure 4.7 Turbidity currents: diagram illustrating the way in which turbidity currents are formed.

and subsequent redeposition by the wind. The grains in such sandstones may be beautifully rounded by innumerable mid-air collisions with each other. Dune bedding is a large-scale form of cross bedding produced by down-wind migration of sand dunes in a way similar to that in which cross bedding is produced under water. Aeolian sandstones often have a warm red-brown colour produced by the deposition of red iron oxide (haematite, p. 55) from solution in water percolating through the spaces between the sand grains.

Argillaceous rocks

The finest-grained detrital sediments are made largely of new minerals such as clay minerals formed during chemical weathering. They are generally deposited as mud (Fig. 1.4) or as the more sticky but still plastic **clay** which is of great importance to man as the principal material used in making bricks and tiles. The squeezing-out of water during diagenesis converts mud and clay to dry brittle rocks. **Mudstone** is a massive argillaceous rock of this kind while **shale** shows a marked tendency to split parallel to the bedding, resulting from the parallel arrangement of the flaky clay minerals in it. All these materials are so fine-grained that their minerals can scarcely be distinguished even with a microscope.

Argillaceous rocks accumulate in the sheltered waters of bays, lagoons and creeks, and in the open sea at depths too great to be reached by strong waves. Because they are deposited in fairly still waters, argillaceous sediments seldom show cross bedding and are most often characterised by a fine parallel lamination; they may contain whole, well-preserved fossils. After diagenesis, the colour of shales and mudstones depends on the presence of iron compounds (which impart a greenish-grey or more rarely red or purple tinge) and on the abundance of organic material which gives a grey or black colour. Because bays, lagoons and creeks provide suitable habitats for the growth of microscopic plants and animals, decayed organic matter is often deposited along with clay minerals. Mud containing a large proportion of organic debris is black and smelly and on burial is converted to **black shale** containing carbon and pyrite (iron sulphide, p. 90), formed by bacterial and chemical reactions. Among the byproducts of these reactions are liquid or gaseous **hydrocarbons** (compounds of the elements hydrogen and carbon) which are the raw materials of oil and natural gas (p. 33).

The chemical–organic rocks

The rocks which are formed through the agency of animals and plants, or by direct precipitation of dissolved matter, are so diverse both in composition and in appearance that it is difficult to make any generalisations about them. We shall deal with the main groups in turn; details are given in Table 9, (p. 67).

Carbonate rocks

Limestones are made largely of calcium carbonate, $CaCO_3$, usually in the form of **calcite**. They are of many kinds. **Organic limestones** are composed of fragments of calcareous shells, corals, echinoderm plates and so on. They often contain abundant fossils (Fig. 4.8), although much of their substance consists either of finely broken material or of microscopic shells. **Chalk** is a soft white limestone made almost entirely of the skeletons of minute organisms. **Chemical limestones** are formed by precipitation of $CaCO_3$ from sea water. The most remarkable of these are **oölites** in which the carbonate is arranged in little spheres to give a creamy-white rock looking like fish roe. Each sphere has been formed by the precipitation of calcium carbonate, layer by layer, around tiny grains or chips of shell lying on the sea floor (Fig. 4.9).

Limestones are formed in seas and lakes where detrital sediment is scarce, particularly in warm shallow seas where shellfish and/or corals flourish

Figure 4.8 A fossiliferous limestone. The most common fossils in this rock are the stalks of crinoids ('sea lilies').

THE CHEMICAL–ORGANIC ROCKS

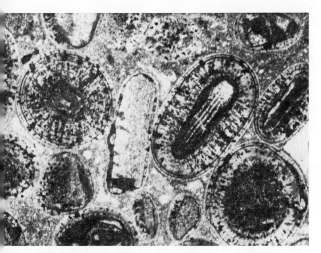

Figure 4.9 An oölitic limestone, seen under the microscope and magnified about 25 times. The oöliths have been cut open to reveal the concentric layering.

and where the water contains abundant calcium carbonate in solution. The clear warm waters off the Bahamas and the Great Barrier Reef off northeastern Australia are two places where limestones are accumulating at the present day.

The abundance of calcite in most limestones makes the rocks easy to recognise because they have many of the same properties as the mineral itself (p. 55). Limestones are usually white, cream or pale grey, and often obviously crystalline. They are rather soft and can be scratched with a knife. Calcite reacts with dilute acid, and most limestones therefore 'fizz' when a drop of acid is spilt on them. **Dolomitic limestones**, which contain calcium magnesium carbonate (dolomite, p. 20), however, may not respond to dilute acid as this mineral is less reactive than calcite. Limestones are important as building stones, and fossiliferous limestones are often used for ornamental facings and floor pavings. Some of the most important British building stones are the Jurassic limestones of the Cotswolds, including the oölitic 'Bath Stone', which were used in many of the historic buildings of Bath and Oxford. Limestone is also the source of lime, which is used in the chemical industry, as a fertiliser and as a constituent of cement.

Siliceous rocks
Silica (SiO_2) is among the commonest natural compounds, occurring not only as quartz but also as a variety of other minerals. Although apparently insoluble, silica will dissolve slowly in waters charged with various solvents, and may be precipitated again on the sea floor during diagenesis. **Flint** and **chert** formed in this way consist of very fine-grained quartz in association with hydrated forms of silica. Flint, which is black when fresh but usually weathers with a white crust, forms nodules or concretions in chalk. It breaks with smoothly curved **(conchoidal)** fractures, and because it is hard and gives a good cutting edge it was much valued by Stone Age man for tools and weapons.

Ferruginous rocks
Ferruginous sediments are chemical deposits in which iron oxides such as magnetite and haematite (p. 55) or iron silicates or carbonates are associated either with chert (see above) or with clayey sediments. They can be easily recognised by their high density and by the rusty staining that results from weathering. **Sedimentary iron ores**, especially the banded cherty varieties, are the major source of iron for industry. In Britain, Jurassic iron ores in Yorkshire, Lincolnshire and Northamptonshire were the chief sources of iron in the 19th century. Most of Britain's iron ore now comes from abroad.

Evaporites
A very interesting group of sedimentary rocks is formed in desert climates where the heat of the sun causes sea water to evaporate, leaving its dissolved salts behind. Shallow closed seas, which are not replenished from the open ocean, may deposit layers of **salt** on their floors as their waters become saturated. Elsewhere the sea water that permeates sediments at the margins of the sea may also be concentrated by evaporation until salts are deposited in the pore spaces of these sediments. Evaporites are formed today by this mechanism around the shores of the Persian Gulf. **Halite** or **common salt** (NaCl, p. 55) is the principal mineral of evaporites formed from sea water. Other minerals are **gypsum** ($CaSO_4 2H_2O$, p. 56) and a number of **potassium and magnesium salts**. All these are important to man: the potash salts as fertilisers, the gypsum for the manufacture of plaster used in building, the refined salt for cooking and all evaporites in various chemical manufacturing processes. Rock salt, or halite rock, is a soft crystalline substance, white or pink when pure but often stained brown by impurities. It is used in winter to treat icy roads and is often stored by local authorities for this purpose.

30 THE SEDIMENTARY ROCKS

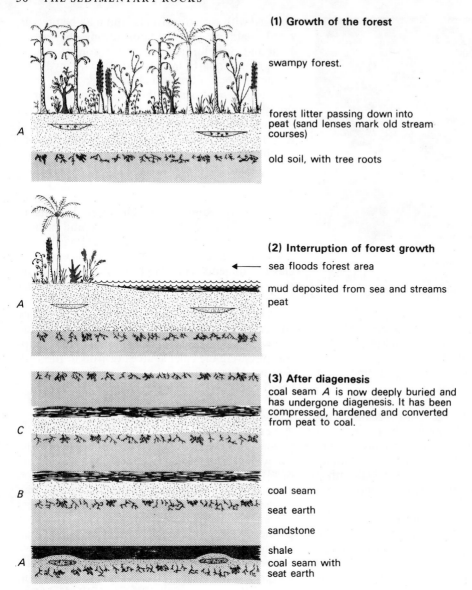

Figure 4.10 The origin of coal seams.

Carbonaceous rocks, especially coal
Along the lower reaches of the Amazon River in South America, dense forests threaded by stagnant streams cover an area of thousands of square kilometres. Dead leaves, branches and seeds which pile up year after year on the forest floor are rotted down to form a soft oozy substance made almost entirely of disintegrating vegetable matter. This forest litter is the **raw material of coal**. When first formed, it is saturated with water and contains many easily recognisable scraps of plants. As it is buried by later accumulations, its water is gradually squeezed out, its plant matter is impregnated with gelatinous organic compounds, and it is gradually converted into denser and more compact **peat**.

You will notice that, unlike most of the sedimentary rocks described above, the raw materials of coal are not transported to a new site before being deposited – they accumulate directly on the forest floor (Fig. 4.10). Indeed, it is sometimes possible to recognise below a coal seam the **old soil** in which the forest trees grew. This soil was penetrated by the tree roots that may still be found in fossil form in the hardened soil underlying coal seams in mines. The fossil soil (which miners call **'seat earth'**) and the **coal seam**, together with the rocks associated with them, tell us something about a period of forest growth long ago.

Coal seams are usually covered by beds of shale which in turn are covered by sandstones. The upward succession of seat earth, coal, shale and sandstone is seen over and over again in coal-bearing sedimentary successions in Britain, and provides an example of a **cyclothem** or **sedimentary rhythm**, which can be interpreted in terms of *changes in the environment of deposition*. The coal seam itself records a long period of forest growth. The overlying shale records a time when the forest was swamped by waters carrying mud and silt; fossil shells show that in some instances the muds were deposited by the sea, and suggest that the ending of the period of forest growth was due to subsidence of the forest floor (Fig. 4.10, stage 2). The overlying sandstones record a period when streams depositing sand and silt gradually built up the land surface to a level at which soils could develop and a new cycle of forest growth could begin.

As each bed of peat in an area such as that described above is buried beneath younger material, it becomes gradually darker, harder and heavier as water and volatile substances are squeezed out. The end-product is a coal seam, from one-fifth to one-fifteenth of the original thickness, which has a very high content of carbon derived from the organic compounds of the original plant matter. The carbon of coal, when heated in the presence of oxygen, is converted to carbon dioxide, CO_2, with the emission of heat. This is what happens when coal burns and is the basis of its value as a fuel. The coal used in Britain comes from **coal measures** – sedimentary piles containing coal seams – deposited some 300 million years ago during the Carboniferous Period, and is derived largely from the debris of swampy forests. These coal measures are preserved at depths convenient for mining in South Wales, Yorkshire and Nottinghamshire, Durham and Central Scotland (Fig. 4.11). The coalfields of these and other regions have been mined for generations and until recently provided most of the fuel used in Britain.

Coal, once such a familiar material, is a black dirty rock with low relative density, usually breaking into rectangular slabs. The original plant matter is so much altered by diagenesis that it is seldom recognisable, though an occasional bedding plane may show a frond or a scrap of wood. Much of the coal is structureless and dull, but occasional layers of solidified gelatinous material show black and shiny surfaces.

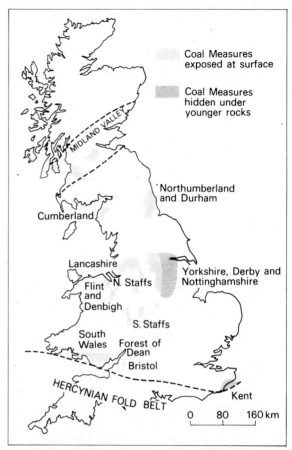

Figure 4.11 The coalfields of Britain. The shaded areas mark the regions where Coal Measures are known to occur at or below the land surface.

Water in sedimentary rocks

Water is always present in larger or smaller amounts in the pores and spaces of sediments. **Connate waters** represent the trapped portions of the sea, river or lake waters in which the sediments were laid down. These may be mingled with **surface water** contributed by rain and with **juvenile waters** derived from volcanic sources. Water may flow through pore spaces and discontinuities not only in sedimentary but also in igneous and metamorphic

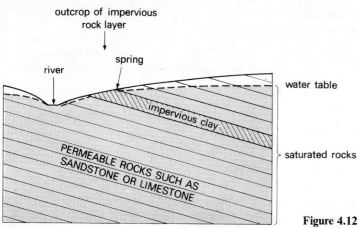

Figure 4.12 The water table.

rocks. As it moves, it carries soluble substances such as calcium bicarbonate or silica which may be deposited later on as a **cement** between the grains (p. 25) or in the form of nodules or **concretions**. Migrating waters may also concentrate and enrich deposits of metals.

Water supply
Fresh water for human use comes partly from rivers and lakes, and partly from underground sources. When rain falls, some of the water evaporates again, or is absorbed by plants; some runs away to join the rivers; and the remainder soaks down to become **groundwater** in the fractures and pore spaces of the rocks. When all pore spaces are filled with water, the rocks are **saturated** and can absorb no more water. Water then seeps out at the surface in hollows and valleys to form lakes, bogs and marshes; **springs** mark the points at which excess water flows out of the saturated ground (Fig. 4.12). After periods of dry weather, on the other hand, the surface layers of soil and rock dry out, air is sucked down into the pore spaces and the seepage to springs and swamps diminishes.

The top of the **zone of saturated rocks** is called the **water table** (Fig. 4.12). It rises and falls with the seasons, but is usually close to the land surface in valleys and some way below the surface in hills. Layers of impervious rock such as clay obstruct the flow of groundwater, and springs may appear above the outcrop of such rocks (Fig. 4.12). The water table marks the level at which water stands in a well and the descent of the water table in dry weather may cause shallow wells to run dry. In England, where the rainfall is high, the water table seldom falls very far beneath the surface, but in hot deserts such as the Sahara of North Africa, where months or years may pass without rain, the water table often lies far below the surface, and both wells and rivers run dry for long periods. In such climates, river water flowing after storms often has to be stored in reservoirs constructed by damming the river for use in dry periods.

The most reliable supplies of **underground water** come from layers of **permeable rocks** – that is, rocks in which the pore spaces are large enough to allow water to flow through them or which are traversed by many fractures or joints. Rocks that can hold and transmit water in this way are called **aquifers**. They are usually, though not always, sedimentary rocks, the most important types being sandstones and limestones (Fig. 4.13). A large part of south-east England is underlain by the soft white limestone known as the Chalk which transmits water through joints. The Chalk is the main aquifer supplying water to London and the home counties (Fig. 4.13). In the Midlands, the red-brown sandstones deposited during the Triassic period are the principal aquifers. Where an aquifer overlain by an impermeable rock, such as clay or shale, is bent downward into a saucer-shaped structure, water is sealed into it under hydrostatic pressure and will rise in wells without being pumped. The Chalk aquifer under London (Fig. 4.13) forms a structure of this kind – an **artesian basin** – and wells sunk to the Chalk formerly produced water which spouted from the surface. The fountains of Trafalgar Square used to flow naturally until excessive pumping reduced the water pressure in the Chalk. Water derived from the Chalk and other limestones is said to be 'hard' because it carries dissolved seas. In ordinary circumstances organic matter is destroyed by bacteria, but where the circulation of water is poor and the oxygen supply is limited, decay is arrested at a stage when many compounds of carbon and hydrogen remain. These **hydrocarbons** occur in mixtures which may be almost solid (**bitumen, asphalt**), liquid (**oil**) or gaseous (**natural gas**). The solids are generally almost black, the liquids yelllowish with a remarkable iridescence and a disagreeable smell.

Oil and gas are formed mainly in argillaceous sediments (p. 28). During diagenesis, liquid hydrocarbons are squeezed out of these source rocks. They move through pore spaces and along bedding planes and fractures, always tending to migrate upwards because they are lighter than the water already filling the pore spaces. If nothing stops them, they eventually reach the Earth's surface and are lost. However, an impervious rock layer such as a clay, through which liquids cannot easily pass, may provide a trap beneath which ascending oil or gas is held. Such an accumulation of oil in pore spaces may be tapped by boreholes drilled from the surface and the gas and oil can be drawn off for commercial use (Fig. 4.14). Oil and gas are extracted from the rocks in enormous quantities and burned as fuel. **Petroleum** is a refined oil used in motor-cars; other grades are used in aeroplanes, for heating purposes and to provide the basis for the manufacture of plastics and many other products. Natural gas can be piped almost directly from the oilfield for use in domestic gas-cookers and so on.

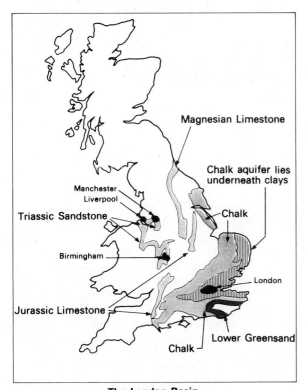

Figure 4.13 Underground water supply. The map shows the positions of important aquifers beneath the land surface. The section (below) shows the arrangement of the Chalk aquifer in the artesian basin under London.

calcium bicarbonate, which reacts with soap to produce an insoluble scum, and deposits calcium carbonate 'scale' when boiled. Water derived from sandstones on the other hand is naturally soft and makes a good lather with soap.

Petroleum, oil and gas

Natural **bitumens** and **oils** are derived from plant matter entombed in sediments. Unlike coals, their source is thought to be the microscopic organisms that flourish in the surface water of lakes and

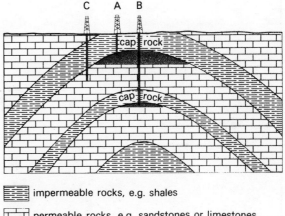

Figure 4.14 Drilling for oil. Wells A and B have struck oil but well C is dry.

Chapter 5

Volcanic Activity and the Igneous Rocks

Igneous rocks are those formed by the solidification of molten rock or magma (p. 15). The temperatures at which most rocks melt are at least eight times higher than the boiling point of water (i.e. 800°C or above), and large volumes of magma can therefore only be formed at depths of many kilometres, where the Earth's internal heat raises temperatures to these levels. As we have seen already (Fig. 2.3, and p. 15), some magma solidifies within the Earth's crust to form **intrusive igneous rocks**: the remainder erupts from volcanoes at the surface to form **extrusive igneous rocks**. **Volcanic eruptions** can be observed, from a safe distance, in the Hawaiian Islands of the Pacific Ocean and in many other active volcanoes (Fig. 5.1). Intrusive igneous rocks, on the other hand, cannot be seen in course of formation, because they are produced from magma that solidifies beneath the Earth's surface, where humans cannot go. Intrusive igneous rocks such as the granite illustrated in Fig. 3.1 are exposed at the surface only after earth movements have raised them above sea level and erosion has worn away the rocks that originally covered them. It is therefore best to begin with the volcanic rocks, about which more is known.

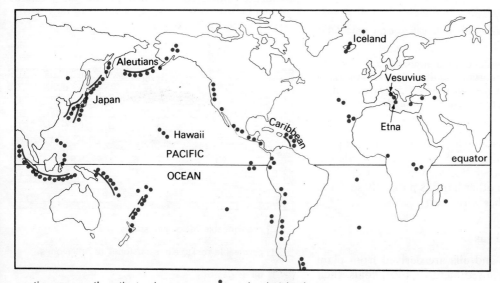

• active or recently extinct volcanoes   ⁙ volcanic island arcs

Figure 5.1 The distribution of active or recently extinct volcanoes and the sites of volcanic island arcs.

Volcanoes and their distribution

Volcanoes develop where magma escapes at the Earth's surface. The liquid magma which flows out from the mouth or vent of a volcano is called **lava** and the rocks formed by the cooling of this liquid are known as lavas. **Volcanic gases** are also emitted from many vents and, as we have seen (p. 10), these gases may escape with such violence that clots of magma and fragments of solid rock are carried up with them into the air. The rocks formed as these materials fall back to Earth are called **pyroclastic rocks** (the word means 'broken by fire').

A map which marks the position of active or recently extinct volcanoes throughout the world (Fig. 5.1) shows that their distribution is very uneven. Many such volcanoes are found around the shores of the Pacific Ocean; the northern and western sides of the Pacific, indeed, are bordered by volcanic islands, many of which are strung out along curved lines such as those that form the Aleutian and Japanese islands. These **volcanic island arcs** have a special geological significance, which will be mentioned in Chapter 9. Outside the **circum-Pacific zone**, concentrations of active volcanoes are seen in the Mediterranean region, in East Africa, and in many groups of oceanic islands which represent the tops of submarine volcanoes. On the other hand, large parts of the continents, including Britain and most of Europe, have no active volcanoes of any kind.

If you compare Figure 5.1 with Figure 2.5 (p. 17), you will see that most of the active volcanoes are found in areas which are subject to earthquakes. This suggests that the melting of rocks to give the magmas which feed volcanic vents takes place mainly in mobile parts of the Earth. Melting is of course dependent on temperature, and many independent lines of evidence show that temperatures within the Earth are generally higher in mobile belts than they are in stable regions.

Volcanic activity

Volcanic eruptions vary in character and in violence. Most are of little danger to man, but some large volcanoes have erupted with great violence in densely populated regions such as Indonesia and the Caribbean, spreading destruction over the surrounding land and causing heavy loss of life. Major eruptions of this kind are very rare and a volcano may remain dormant for centuries before coming to life with catastrophic violence. There is, therefore, a great need for us to understand the workings of volcanoes in order that accurate predictions of future eruptions can be made. Volcanologists classify eruptions according to several features which are listed in Table 4.

The first division depends on the route taken by magma as it rises through the crust (Fig. 5.2). Magma rising in a long crack or fissure erupts at many points along a line. Lava escaping during **fissure eruptions** of this kind does not build up a definite cone, but flows out to form a flattish **lava plateau**. Magma solidifying in the crack from which eruption takes place forms a thin vertical sheet or **dyke** (Figs. 5.2a, 5.3). Magma that rises in a chimney-like duct gives rise to a **central eruption** and piles up lava and pyroclastic rocks around the **volcanic vent** to produce a conical mountain with a central hollow or **crater** (Fig. 5.2b). Magma solidifying in the duct forms a vertical cylindrical body, which is called a **volcanic plug** or **neck**. The classic shape of a **volcanic cone**, like Mount Fuji in Japan, is usually modified by the development of subsidiary cones and fissures on its flanks, by the destructive effects of explosions (see below) and, of course, by river erosion.

The effects of each eruption depend on the properties of the magma involved and, in particular, on its **viscosity** or ease of flow. You can gain some idea of the influence of viscosity by comparing the ways in which warm and very cold treacle flows off a spoon. Warm treacle has a low viscosity and is 'runny', moving quickly and smoothly in thin streams. Treacle from the refrigerator has a higher viscosity and flows more slowly, in thick streams trapping accidental air bubbles. The types of eruption classified under item 3, Table 4, are related to magmas arranged in order of increasing viscosity from (a) to (e).

Magmas with relatively low viscosity are responsible for most submarine eruptions, most fissure eruptions, and for Hawaiian and Strombolian eruptions from central volcanoes. The lava flows smoothly from the vent and spreads out into a thin, extensive tongue or sheet. The intervals between eruptions are short, and consequently the vent does not become entirely choked by solidified lava. Most important of all, dissolved gases – water vapour, carbon dioxide, sulphur dioxide and others – carried by the magma can escape easily to the air, like gas escaping from fizzy lemonade. The central volcanoes characterised by eruptions of this kind have a broad, low

36 VOLCANIC ACTIVITY AND THE IGNEOUS ROCKS

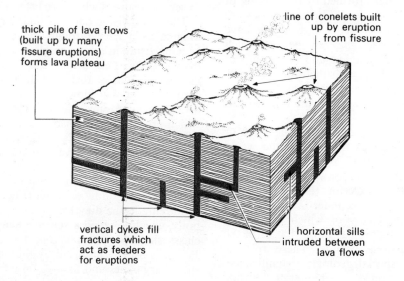

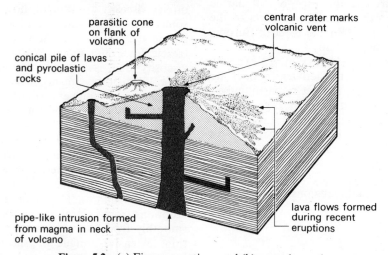

Figure 5.2 (a) Fissure eruptions and (b) central eruptions.

Table 4 Classification of volcanic eruptions.

1. *Shape of vent:* (a) *Fissure eruption:* lava escapes from elongated crack (Fig. 5.2a)
   (b) *Central eruption:* lava escapes from circular vent (Fig. 5.2b)
2. *Site of eruption:* (a) Submarine: lava flows out onto sea floor
   (b) Subaerial: lava flows out on land
3. *Type of eruption:* dependent on viscosity of magma and volume of dissolved volcanic gas.

(a) *Hawaiian:* frequent but quiet eruptions, magma of low viscosity; much lava, gases escape quietly, therefore few pyroclastic rocks.
(b) *Strombolian:* frequent eruptions producing mainly lava; gas escapes with small explosions and most fragments thrown up by it fall back inside crater.
(c) *Vulcanian:* longer intervals between eruptions allow lava to solidify in volcanic neck; trapped gases cause strong explosions which shatter vent walls and throw rock fragments out of crater; both lavas and pyroclastic rocks produced.
(d) *Vesuvian:* intervals between eruptions usually more than 25 years; trapped gases escape in violent up-rush carrying fragments and dust; fine volcanic ash may fall up to 10 km from vent; more pyroclastic rocks than lavas.
(e) *Plinian:* most violent type, associated with viscous magma; rare but violent eruptions may destroy existing volcanic cone; pyroclastic rocks abundant and widespread; lava may fail to escape forming a dome in the vent.
(f) *Peléan:* intervals between eruptions up to many centuries, eruptions involve magma rich in gases; fast-moving gas cloud charged with magma and volcanic dust flows out from vent as 'glowing cloud' or **nuée ardente** and travels many kilometres before solidifying; abundant pyroclastic rocks; lava usually fails to escape and forms a dome inside the vent.

Figure 5.3 A dyke of basic igneous rock, Isle of Arran: the hammer is propped against the dyke. (Crown copyright)

More viscous magmas move slowly, tend to solidify in the neck of the volcano and, when they do escape, form short, thick flows close to the vent. Gas coming out of solution becomes trapped in the pasty lava; gas pressure then builds up to a level which is sufficient to shatter the lava and the walls of the crater. The violent uprush of gas carries rock fragments and dust high into the air from which they fall back to form pyroclastic rocks. For this reason, Vesuvian and Plinian eruptions produce relatively little lava but large

profile well illustrated by those of the Hawaiian islands (Fig. 5.4), and are called **shield volcanoes**.

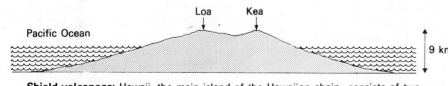

Shield volcanoes: Hawaii, the main island of the Hawaiian chain, consists of two coalescing shield volcanoes.

Figure 5.4 Shield volcanoes. The figure shows the low cones of two coalescing volcanoes in the Hawaiian Islands.

38 VOLCANIC ACTIVITY AND THE IGNEOUS ROCKS

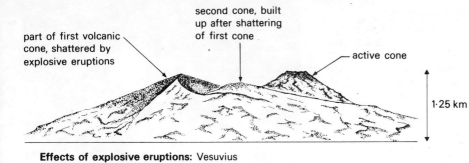

Figure 5.5 Vesuvius, seen from the sea. The broken and eroded rim of an old cone is on the left and two younger cones can be seen within it.

volumes of pyroclastic rocks. These are destructive eruptions, because the finer ash settles out over wide areas, killing crops and crushing buildings, while the accompanying gases can asphyxiate humans and domestic animals. Plinian eruptions are called after a Roman youth Pliny who lived through the eruption of Vesuvius, on the Bay of Naples in AD 79 and described its effects in letters to the historian Tacitus. During this eruption the town of Pompeii, which was downwind from the volcano, was smothered in hot ash and some 2000 of its inhabitants were killed. The lopsided shape of Vesuvius today (Fig. 5.5) has been built up during many subsequent eruptions; the old volcanic cone has been shattered by explosions, and younger ash and lava cones have grown up inside it.

The most catastrophic of all eruptions are those of the Peléan type, named after Mont Pelée in the Caribbean island arc which erupted in this fashion in 1902. Escaping gas acts as a lubricant to a dense, glowing cloud of particles and magma droplets which flows out at white heat close to the land surface, burning and suffocating everything in its path and finally coming to rest as a solid sheet in which fragments are flattened and welded by intense heat. The eruption of such glowing clouds (the French phrase **'nuées ardentes'** is used for them) is associated with the rise of magma so viscous that it squeezes up in the volcanic neck like toothpaste, forming a volcanic dome or spine which is soon destroyed by gas blasting. The emptying of a magma chamber by the eruption of nuées ardentes or from other causes sometimes leads to the collapse of the whole top of the volcano, forming a circular depression several kilometres across, known as a **caldera**.

Lavas and pyroclastic rocks

The forms of many lavas give a vivid impression of the motion of the parent magma. A ropy or blocky surface records the wrinkling and ultimate breaking up of a solidifying crust pulled along by the hotter and more fluid magma beneath (Fig. 5.6). **Vesicles,** or voids in the lava, represent gas bubbles frozen in and often pulled out of shape during solidification (Fig. 5.7). **Pumice** is an extremely light porous rock formed by solidification of lava froth. **Pillow lavas**, looking like piles of half-filled sacks, are formed from magma oozing out into water or mud (Fig. 5.7).

The pyroclastic rocks are formed in volcanic necks and on the land surface as a result of explosive volcanic eruptions powered by escaping water vapour, carbon dioxide and volcanic gases generally. The tremendous forces released – many times those generated by a hydrogen bomb – shatter the wall rocks and sometimes the solidifying magma itself. A jumble of coarse angular fragments remains close to the volcanic vent as an **agglomerate**. The finer **volcanic ash** settles out over large areas to form layers of fine particles which compact to produce a bedded **tuff**. Occasionally, the spontaneous outrush of gas causes a volcano to 'boil over' in a white-hot nuée ardente which solidifies as a massive sheet of rock known as an **ignimbrite**.

The textures of lavas reflect the rapid cooling that results from exposure to air or water. Most are fine grained and must be examined with a lens or microscope. The most rapidly cooled lavas are sometimes not crystalline at all. The ground-mass is a natural **glass** or super-cooled liquid which fractures like bottle-glass and is translucent in thin splinters. Most lavas, however, are made up of a

Figure 5.6 A basic lava flow showing a ropy top due to wrinkling of the surface during flow.

Figure 5.7 Pillow lavas from the Ballantrae area, Scotland. Erosion has revealed the internal structure of the lava pillows which show concentric layers of vesicular and non-vesicular lava. (Crown copyright)

groundmass of minute granular or elongated minerals interlocking with each other. Large and well-formed crystals of feldspar or other minerals may be scattered through this groundmass to produce a **porphyritic texture** (Fig. 5.10, p. 43). Large **phenocrysts** of this kind began their growth before the lava reached the surface and began to cool.

Igneous intrusions

The shapes of the intrusive bodies of magma which squeeze themselves into spaces within the Earth's crust depend on the structure of the crust itself and on the stresses acting on it. Two kinds of igneous intrusion have been mentioned already because they are formed in connection with volcanic eruptions. Dykes are thin vertical sheets, some of which 'feed' fissure eruptions (Figs 5.2, 5.3); these fissure fillings, often only a few metres thick, tend to occur in swarms filling parallel fractures; such **dyke swarms** are often intruded in the gaps opened by stretching of the crust. Volcanic plugs or necks are the feeders of central volcanoes (Fig. 5.2); they are roughly circular or oval in plan and usually no more than a kilometre or so in diameter. Plugs and dykes are often associated with flat-lying intrusive sheets or **sills**, most of which were formed by magma squeezing its way along the bedding planes in sedimentary rocks (Fig. 5.2). Larger intrusive bodies (Fig. 5.8) may form steep-sided masses with domed or arched roofs. Those of about 10–50 km diameter are called **stocks** or **bosses**, and still larger intrusions are called **batholiths**. The intrusion of several successive pulses of magma up the same duct may produce a **ring complex**, showing in plan several ring-shaped bodies, one within the other. Finally, large magma bodies intruded in horizontal planes tend to form lenticular or sagging sheets known as laccoliths and lopoliths respectively.

All intrusive igneous rocks share some general characters which help to distinguish them from the extrusive rocks. I shall mention these general characters here, leaving detailed descriptions for a later section (p. 46). The intrusive magma which comes to rest within the crust loses heat only rather

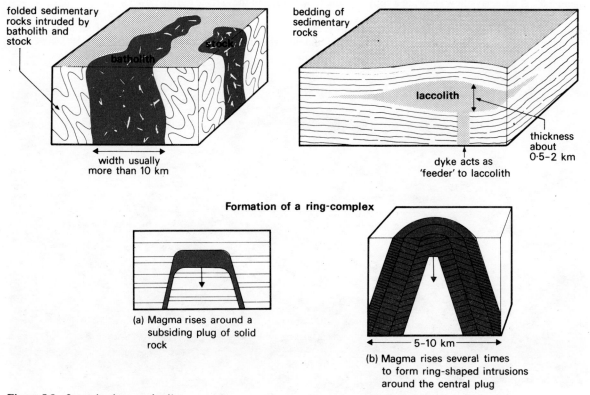

Figure 5.8 Intrusive igneous bodies.

slowly by conduction. Slow cooling allows time for the growth of crystals and intrusive rocks are therefore generally entirely crystalline. The rocks of large, slow-cooling intrusive bodies tend to be **coarse grained**, while those of dykes, sills and small plugs are usually **medium grained** or **fine grained**. The igneous rocks of large intrusions are called **plutonic** rocks after Pluto, the king of the underworld in Greek mythology.

The coarse-grained plutonic rocks (average grain-diameter is over 5 mm) are obviously crystalline, and individual minerals in them may be fairly easily identified with the naked eye. The fine-grained igneous rocks (diameter less than 1 mm) do not look cleanly crystalline to the naked eye and may have a dull surface when broken. Their crystalline nature, however, is usually apparent when they are examined with a hand lens, and is always clear when a thin slice of rock (ground down until it becomes transparent) is looked at under the microscope (Fig. 5.10).

At the borders of intrusive igneous bodies where hot magma came in contact with colder crustal rocks, a **chilled margin** made of finer-grained rock is often seen. Large intrusions may be surrounded by a **contact aureole** in which the 'country rocks' have been baked or recrystallised by the heat of the intrusive magma (see p. 46). Veins of igneous rock may penetrate the country rocks and fragments of the country rock picked up from the walls may be seen in the intrusion. Such fragments are known as **xenoliths** from the Greek word *xenos* = a stranger.

Classification of igneous rocks

The minerals that crystallise from cooling magmas are those which contain the elements present in the magma. The **mineral composition** of any igneous rock is therefore closely related to its **chemical composition**, and both characteristics can be used to classify the igneous rocks. Since the chemical composition cannot be accurately determined without analysis, most geologists adopt a rough-and-ready classification based, as far as possible, on the presence or absence of certain minerals which provide an approximate guide to chemical composition.

The principal chemical component of all igneous rocks is silica (SiO_2) which forms from about 40% to over 75% of their total weight and which is present in all silicates as well as in quartz. The igneous rocks are usually grouped *according to silica-percentage* as follows:

		% SiO_2
(a)	acid rocks	>66
(b)	intermediate rocks	52–66
(c)	basic rocks	45–51
(d)	ultrabasic rocks	<45

To these chemical groups must be added a fifth, that of the **alkaline rocks**, in which the silica percentage ranges from basic to acid and the percentage of alkalies (sodium, Na and potassium, K) is higher than in the four groups listed above.

The next step in classification depends on grouping the rocks *according to their mineral composition* as is illustrated in Figure 5.9. Table 2 on page 20 which gives the compositions of the silicate minerals shows that ultrabasic and basic rocks are rich in minerals which contain iron, magnesium and calcium (collectively called **ferromagnesian minerals**). Acid rocks are rich in quartz (SiO_2) and feldspars (the **felsic minerals**) which contain silica, alkalies (sodium, potassium) and aluminium. By identifying the most important minerals and making a rough estimate of their proportions, it is possible to guess at the composition of an igneous rock. This method of identification, of course, breaks down when the rock is glassy or when it is so fine grained that individual minerals cannot be identified without a microscope; many fine-grained lavas

Table 5 Classification of igneous rocks.

In increasing grain size	*Ultrabasic*	*Basic*	*Intermediate*	*Acid*
fine-grained or glassy	(no common rocks)	basalt	andesite	rhyolite obsidian pitchstone
fine- or medium-grained	(no common rocks)	dolerite	porphyrite	porphyry, microgranite
medium- or coarse-grained (plutonic rocks)	peridotite	gabbro	diorite	granite

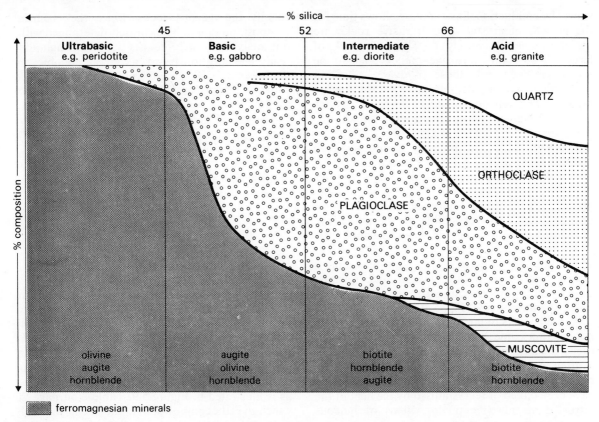

Figure 5.9 The chief minerals of igneous rocks. Orthoclase is potassium feldspar, plagioclase is sodium-calcium feldspar. Biotite is iron-magnesium mica, muscovite is aluminium-rich mica.

therefore cannot be accurately identified in hand specimen.

The final stage in the naming of an igneous rock depends on its **texture**. The main divisions are drawn between the lavas and associated rocks which have fine-grained or glassy textures, the medium-grained rocks formed in small intrusive bodies and the coarse-grained (plutonic) rocks of large intrusions. The names of rocks in each category are given in Table 5 which summarises the classification used in this book. In the paragraphs that follow, the principal varieties are described briefly. Table 10 on page 68 lists the characters that distinguish each type.

Ultrabasic and **basic rocks**, as Figure 5.9 shows, contain pyroxenes associated with olivine or with plagioclase feldspar. They can usually be identified by their high relative density (up to 3·3, as compared with granite 2·7–2·6) and by their colour. Since most ferromagnesian minerals are deeply coloured, rocks of basic or ultrabasic composition, which are wholly or largely made of these minerals, appear black, dark grey or dark green on fresh surfaces. The **peridotites** and **gabbros** are coarse grained, the peridotites being entirely green or black, the gabbros often being mottled by patches of white plagioclase. **Dolerites** are medium grained, black, green or grey rocks. **Basalts** are dark and fine grained but often porphyritic (Fig. 5.10). All these types are more or less **massive**; that is to say, the minerals do not have a strong parallel arrangement and the rocks present much the same appearance when looked at on three mutually perpendicular surfaces.

Basalts (Fig. 5.10) are by far the most abundant lavas found on the Earth's crust and appear to be formed from magmas that represent the commonest products of melting of material from the mantle. Basalts are erupted from fissures along the oceanic ridge system (p. 63) and make most of the oceanic

feldspar phenocrysts

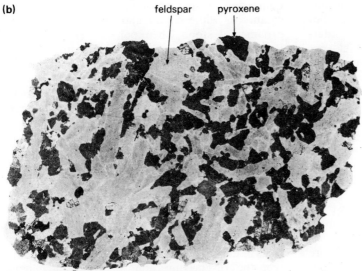

feldspar pyroxene

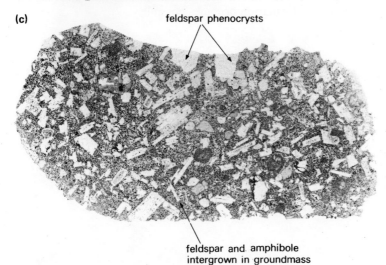

feldspar phenocrysts

feldspar and amphibole intergrown in groundmass

Figure 5.10 Thin sections of three igneous rocks photographed under a microscope, enlarged about 10 times. (a) Porphyritic basalt showing feldspar phenocrysts in a glassy groundmass. (b) Dolerite, a medium-grained basic rock. (c) Porphyritic andesite showing feldspar phenocrysts in a fine crystalline groundmass.

crust. They form perhaps 95% of the vast oceanic volcanoes such as those of Hawaii and also form gigantic lava-plateaux on the continents. The Giant's Causeway of Northern Ireland is a remnant of such a basaltic lava pile. Intrusive basic rocks are found within the crust beneath regions where basalts have been erupted. Dykes and sills are commonly made of dolerite, and larger bodies, such as laccoliths, of gabbro.

The **intermediate rocks** are generally of little importance compared with the basic group. However, **andesitic lavas** – dark grey, brownish or greenish, rocks, fine grained and often porphyritic – are erupted in enormous volumes in volcanic island arcs and in the mountain regions of the **orogenic mobile belts**; indeed they take their name from one such region, the Andes of South America. The abundance of andesites in the zones along which the crust has been compressed (Ch. 10) suggests that their parent magma may be derived from the surplus crustal material crowded down into the hot mantle beneath these belts. Andesites, as intermediate rocks, usually contain plagioclase feldspar and augite or hornblende (Fig. 5.10c). One or both of these minerals may form phenocrysts large enough to be easily identified, but the minerals of the groundmass are seldom recognisable. In **porphyrites**, phenocrysts of plagioclase, augite or hornblende are set in a rather less fine-grained groundmass. The coarse-grained **diorites** are patchy or speckled rocks in which a pale plagioclase feldspar is associated with dark augite or hornblende.

Acid rocks are represented in the crust by relatively small amounts of lavas and by much greater volumes of the coarse-grained type known as granite. They are all characterised, as Figure 5.9 shows, by the abundance of quartz and feldspar and the scarcity of ferromagnesian minerals. The common acid lavas are of several types. **Rhyolites** are pale, very fine-grained rocks often showing contorted flow structures and carrying phenocrysts of quartz or feldspar or both. Two glassy varieties, **obsidian** and **pitchstone** are black, vitreous and shiny when fractured; pitchstone has a resinous lustre. These lavas, often associated with tuffs and ignimbrites (p. 38), occur in small quantities in some basaltic provinces and in larger amounts with andesites in orogenic mountain belts.

Granites are among the most important acid rocks, rivalling basalts in abundance. They occur as large intrusive bodies – ring complexes, stocks or batholiths – and are consequently coarse grained. The principal minerals, quartz, sodic or potassic feldspars and micas (either biotite or muscovite or both) are easily distinguished with the naked eye and large feldspar phenocrysts may reach lengths of several centimetres. Owing to the dominance of quartz and feldspar, granite usually has a pale pink, creamy or white colour and makes a handsome ornamental stone when polished (Fig. 3.1). In Britain, the main granite building stones come from Aberdeen, Shap Fell (Lake District), Dartmoor and the Channel Islands. **Microgranites**, as their name suggests, are finer-grained rocks containing the same minerals as granite. **Quartz-porphyries** are distinguished by the presence of quartz phenocrysts in a fine-grained matrix of quartz and feldspar.

Granites, unlike basalts, *are confined to the continental regions* where they make up a large part of the crust. Their restriction to the continents has led geologists to conclude that part of their substance is derived from the crust itself rather than from the mantle. They are developed in the deep parts of the orogenic mobile belts where, as we have seen, the crust becomes abnormally hot and magma formation becomes possible.

Chapter 6

The Metamorphic Rocks

Environments of metamorphism

The sedimentary and igneous rocks, which are formed at the Earth's surface or in intrusive magmatic bodies within the crust, may be transformed into metamorphic rocks long after the original time of their formation. Their original characters are not as a rule entirely obliterated: for example, bedding may often be preserved in a metamorphosed sedimentary rock (Fig. 6.1) and the large phenocrysts of a porphyritic lava may remain recognisable. The metamorphic rocks are especially interesting as records of geological history since they give evidence of a whole succession of events: the results of deposition of a sediment, of diagenesis and of one or more episodes of metamorphism may all be recorded in a single rock.

The principal factors controlling metamorphism are temperature and pressure. Metamorphic reactions begin at temperatures of a few hundred degrees centigrade and continue up to temperatures of over 800°C, above which large parts of the rocks affected begin to melt. The pressures involved in metamorphism are produced partly by the weight of overlying rocks, in which case they are roughly proportional to the depth in the crust. At depths of 10 km, total pressures are 3000–4000 times greater than atmospheric pressure (= 3–4 kilobars). Near the base of the crust, at 30–40 km depth, they rise to over 10 kb. Compressional forces acting in orogenic belts (p. 17) give rise to **directed stress** whose effects are shown by the fact that metamor-

Figure 6.1 Bedding preserved in a metamorphosed sandstone from Mallaig, Scotland. (Crown copyright)

46 THE METAMORPHIC ROCKS

Figure 6.2 Distortion resulting from earth movements during metamorphism. These fossils trilobites have been squeezed out of shape (about natural size).

phic rocks have often been distorted: the bedding may be crumpled, and fossils may be flattened or squeezed out of shape (Fig. 6.2). The tilting or folding of bedded sedimentary rocks makes it difficult for the geologist to distinguish older beds from the younger beds originally deposited on top of them (p. 14). The original **'way up'** of beds such as those illustrated in Figure 6.1 is sometimes shown by cross bedding or graded bedding, as explained on pages 24 and 27.

The environments in which metamorphism takes place most readily are, naturally, those in which exceptionally high temperatures or directed stresses are generated. In practice, we can group them in two main categories; the environments of **local metamorphism** and those of **regional metamorphism** as shown below

(a) *Environments of local metamorphism:*
  (i) **contact aureoles** surrounding igneous intrusions;
  (ii) **zones of dislocation metamorphism** along faults and zones of movement in the crust.
(b) *Environments of regional metamorphism:* the deeper parts of the orogenic mobile belts.

Contact aureoles

Metamorphism in the contact aureole of an igneous intrusion (p. 41) depends fundamentally on the outflow of heat from the intrusive magma. A small intrusion such as a dyke seldom carries enough heat to do more than bake, harden or discolour the country rocks for a centimetre or so on either side. A large batholith may produce an aureole a kilometre or more in width within which the country rocks are wholly or partly recrystallised. The **metamorphic minerals** developed in the aureole are minerals which are *in equilibrium at the temperatures and pressures attained*. Since the rocks closest to the intrusion are likely to reach higher temperatures than those further away, it is reasonable to expect that the effects of metamorphism will increase progressively towards the contact. In most aureoles, two or more **metamorphic zones** representing different **grades of metamorphism** can be distinguished as illustrated in Figure 6.3.

The outer zone of the aureole is often characterised by **spotted rocks** in which partial recrystallisation has taken place at temperatures of 300–400°C. Metamorphic minerals appear at numerous centres as knots or definite crystals up to a centimetre or so across, while the intervening material remains fine grained. Where the fine-grained groundmass is slaty (p. 47), rocks in this zone form **spotted slates**. The inner zone is characterised

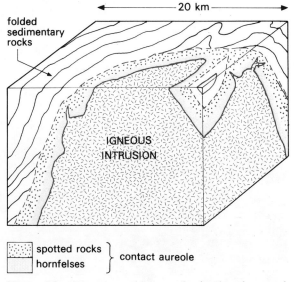

Figure 6.3 A metamorphic aureole developed around a stock of intrusive igneous rock.

by complete recrystallisation, with the production of tough even-grained rocks known as **hornfelses**. Micas are characteristic minerals in many hornfelses derived from argillaceous sediments, whilst hornblende often appears in hornfelsed basic rocks. **Andalusite** (Al_2SiO_5), an aluminium silicate with a distinctive square-sectioned matchstick-like crystal form, is developed in argillaceous rocks which are rich in aluminium.

Intrusive magmas such as granite magma carry dissolved gases which are gradually concentrated during crystallisation near the roof of the intrusion. These **volatile components** of the magma finally escape into the aureole where they may crystallise in fractures as **veins** or may react with the aureole rocks. The veins and reaction zones formed in this way sometimes contain concentrations of copper, zinc and other sulphides or of tin oxide. The mining industry of Cornwall, which was founded before the Romans came to Britain, is based on deposits of tin and copper in and around the aureoles of several granite intrusions (Fig. 7.8).

Dislocation metamorphism

The jerky crustal movements which lead to earthquakes are localised along breaks in the crust called **faults**. In the upper parts of the crust, fault movements tend to shatter the rocks into angular fragments known as **fault breccia**. At greater depths, the rocks along fault zones are often powdered and welded into fine-grained platy rocks in which broken and twisted remnants of earlier minerals may still be seen. These products of dislocation metamorphism are known as **mylonites**.

Regional metamorphism

The orogenic mobile belts where regional metamorphism takes place extend over large areas in which temperatures are abnormally high and compressive stresses often strong. Regional metamorphism in these belts may affect areas many hundreds of kilometres in length. In Britain, for example, most of the Scottish Highlands is built of metamorphic rocks formed in very old orogenic belts which have been deeply eroded (Fig. 6.4).

The phenomenon of metamorphic zoning, which we encountered in the contact aureoles of intrusive bodies, is displayed on a vaster scale in areas of regional metamorphism. We may classify the products in terms of their metamorphic grade into three groups; the low-grade rocks typified by

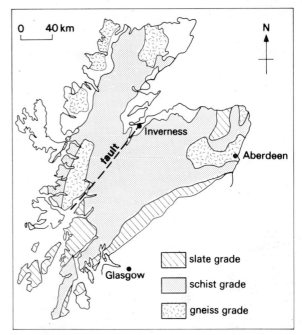

Figure 6.4 Zones of regional metamorphism in the Scottish Highlands.

slates, the medium-grade rocks typified by **schists** and the high-grade rocks typified by **gneisses**. Some of the products, and some characteristic minerals formed at each grade, are mentioned below.

The low-grade rocks

Because the temperatures reached during low-grade metamorphism are only 300–400°C, the grain size of the new minerals developed is generally small. Many low-grade rocks are too fine grained for the groundmass minerals to be seen with the naked eye. Relics of structures and textures of the parent rocks are sometimes still visible and fossils are not always entirely destroyed, though they tend to be strongly distorted (Fig. 6.2).

Slates are the characteristic low-grade rocks derived from argillaceous sediments. The high alumina content of these rocks is expressed by an abundance of white mica and related silicates with flaky habit such as the soft green mineral **chlorite**. The distinctive feature of slates is the alignment of these flaky metamorphic minerals to produce a parallel structure along which the rocks split. This **slaty cleavage** is not always parallel to the original bedding (Fig. 6.5) and is developed in response to compressive orogenic stress. It is often

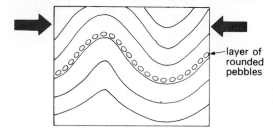

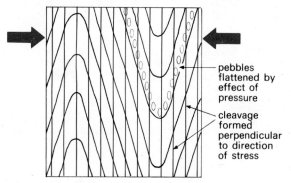

Figure 6.5 Cleavage in a metamorphic rock.

so perfect that slates can be split into layers less than a centimetre in thickness, a property which gives them their importance as roofing material.

The medium-grade rocks
At higher temperatures, the rates of chemical change increase and new minerals are enabled to grow to larger sizes. Most rocks of the middle grades of regional metamorphism present an obviously crystalline appearance and many contain occasional larger minerals (**porphyroblasts**) up to a centimetre or more in diameter. This coarsening of the texture has the effect of obliterating small-scale original features, though larger-scale original structures such as bedding are often well preserved.

The characteristic feature of regionally metamorphosed rocks is the development of new minerals with a roughly parallel arrangement. The parallelism of flaky minerals such as micas or elongated minerals such as hornblendes produces a **schistosity**. Schists with well developed schistosity are common rocks of the medium grades of regional metamorphism. **Mica schists** rich in biotite or muscovite are formed from sedimentary rocks of clayey composition whose high alumina content favours the development of the micas. Other aluminous minerals which may appear in these rocks are **garnets** (p. 54) which often form conspicuous porphyroblasts. **Hornblende schists** are common products of basic igneous rocks in which the high content of calcium, iron and magnesium contributes to the formation of hornblende.

Among other products of regional metamorphism, we may mention quartzites derived from psammitic sedimentary rocks and marbles derived from limestones. Quartzites, as the name implies are rich in quartz and are often pure white in colour. Metamorphic quartzites are often called **metaquartzites** to distinguish them from sedimentary quartzites. The bedding (including cross bedding) may be indicated by parting-planes coated with micas, or by dark lines representing impurities in the original sandstone. **Marbles**, like the limestones which are their parent rocks, are made dominantly of calcite ($CaCO_3$) though they may contain silicate minerals such as amphibole, pyroxene or olivine derived from sandy or clayey impurities. Marbles are pale rocks usually white, cream or pink in colour but sometimes patched or streaked with green or brown impurities; they scratch easily and effervesce with dilute HCl on account of their high calcite content. They form ornamental building materials, used as facing slabs, floor pavings, mantelpieces and so on. Pure, even-grained marbles with few fractures are used for sculpture; one of the most famous statuary marbles is the white Carrara marble used by Italian Renaissance sculptors and seen in monuments in some cemeteries. Because the principal minerals of quartzites and marbles – quartz and calcite respectively – are stable over a wide range of temperatures, these rock types are not very sensitive to variations in metamorphic grade. Low-grade varieties are finer grained and contain minerals such as chlorite in place of hornblende or pyroxene, but otherwise do not differ much from those described above.

High-grade rocks
The characteristic rocks of the highest metamorphic grades are gneisses which show a rather poor parallel structure known as **foliation** in place of the closely spaced schistosity seen at medium grades (Fig. 6.6). Gneisses are coarse-grained rocks in which the felsic and ferromagnesian

REGIONAL METAMORPHISM 49

Figure 6.6 A banded gneiss from the Scottish Highlands in which the foliation has been crumpled by earth movements. (Crown copyright)

minerals of different species are segregated into layers or lenses giving the rocks a streaky appearance. Very few original features survive after high-grade metamorphism.

The minerals formed at high metamorphic grades are those which are stable at temperatures of 600–800°C. All true gneisses contain feldspars which appear in the derivatives of argillaceous rocks, of impure sandy rock and of a wide range of igneous rocks. Biotite gneisses or biotite–garnet gneisses are common derivatives of argillaceous parent rocks, hornblende gneisses of impure calcareous rocks and of basic or intermediate igneous rocks.

A new feature seen in many high-grade gneisses is the appearance of patches, layers and veins of granite mingled with the substance of the metamorphic rock. These mixed rocks or **migmatites** record the first stages of melting in the crust. Granite has a lower melting point than the majority of crustal rocks and hence as temperatures rise during metamorphism, granitic fluids begin to gather in the interstices of the solid rocks. The migmatites lie on the border between the igneous and metamorphic rocks and provide the last link in the cycle of rock-forming processes illustrated in Figure 2.3.

Chapter 7

Properties and Uses of Minerals

Having considered the rocks of the three main classes in Chapters 4, 5 and 6, we must now return to the subject of minerals. As was stated on page 18, minerals are the natural units of which most rocks are made. When you have learned to recognise the rock-forming minerals, you will be able to describe and understand the rocks themselves more completely. First of all, however, we need to consider the properties of minerals in general. You will find it easier to follow the next few paragraphs if you can handle some common minerals such as calcite, fluorspar, quartz or pyrite. Tables 7 and 8 on pages 65 and 66 give details of the common rock-forming minerals.

Examination of many examples of these and other minerals shows that they are not simply irregular grains, but that they have characteristic shapes bounded by many plane surfaces which meet at regular angles. Fluorspar (calcium fluoride, CaF_2), for example, often occurs as cubes, bounded by faces at right angles to each other (Fig. 7.1). **Quartz** (SiO_2) may occur in elongated shapes outlined by six long faces called prism faces and capped by a pyramid of small faces arranged in multiples of three (Fig. 3.2). Regular geometrical forms such as these are called **crystals** and they tell us something about the internal structure of the mineral.

Crystals and crystal symmetry

The growth of any mineral whether silicate or non-silicate takes place in a regular way. Once the first clusters of atoms have been built up, further atoms are added in such a way as to extend the crystal structure according to a regular pattern. Just as a roll of wallpaper may exhibit many repetitions of a small design, so the **atomic lattice** exhibits, in three dimensions, many repetitions of a 'unit cell' in which the atoms are arranged according to a set pattern.

The whole substance of the mineral is thus built according to a single plan and its physical properties are controlled by this plan. The **relative density** of the mineral, its **hardness** when scratched, the **cleavage** (the regular fractures some minerals develop as they break up under a blow) and the rate at which it transmits light or X-rays, all reflect the atomic structure, as does the crystal form. In particular, the regular arrangement of atoms in crystalline minerals determines the symmetry of the crystal forms. You can investigate the idea of **crystal symmetry** by looking at a cubic mineral such as fluorspar or at a plastic or wooden model of such a mineral. Hold the cube up between finger and thumb placed at the centres of opposite faces (Fig. 7.2). Start with one face towards you, then turn the crystal through a right angle. A new face, exactly like the first, comes into view; another turn brings a third and another a fourth

Figure 7.1 Fluorspar: a group of cubic crystals.

OTHER DIAGNOSTIC PROPERTIES

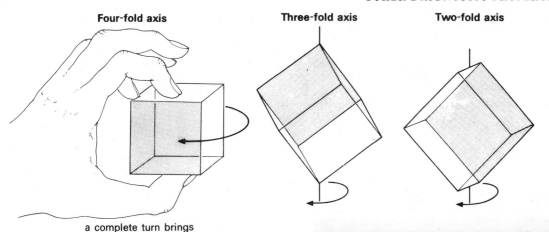

Figure 7.2 Crystal symmetry. The figure illustrates the axis of symmetry of a cube. How many four-fold axes has the cube? How many three-fold and two-fold axes has it? A cubic crystal of fluorspar is shown for comparison.

before the original face comes to the front again. The axis around which you are turning the crystal is an axis of **four-fold symmetry**. If you change your grip to hold the crystal by opposite corners, or by the centres of opposite edges, you can identify axes of **three-fold** and **two-fold symmetry** (Fig. 7.2). Experiments with quartz or calcite crystals (Figs 3.2 and 7.6) or with models of these minerals, will reveal no axes of four-fold symmetry, but will show that three-fold and two-fold axes are present. It is therefore possible to classify minerals according to their crystal symmetry which in turn *depends on their atomic structure*.

The regular structure which results from the arrangement of atoms is built into all crystalline minerals. Although this regularity is often expressed by a geometrically regular crystal form, as is illustrated by the minerals shown in Figures 3.2 and 7.1, perfect crystal forms can develop only where minerals have space to grow without interference. In a cooling magma, many small crystals grow into each other as they enlarge and therefore develop irregular interlocking boundaries. In the granite illustrated in Figure 3.1, for example, the large feldspars have a rectangular outline which corresponds to the common crystal form of the mineral, but their actual boundaries are not smooth crystal faces because, during the last stage of growth, the feldspars had to compete for space with the minerals around them. In clastic sedimentary rocks, grains of quartz and other minerals derived from the erosion of older rocks (p. 26) have been broken and battered during transport to such an extent that no crystal form remains. Many of the minerals that you may see in the classroom or may find for yourself will not be perfect crystals. The symmetry of their atomic structure, of course, remains the same even when it is not expressed by a regular crystal form.

Other diagnostic properties

Composition and lattice structure
The differences between mineral species all depend ultimately on differences in the composition and structure of their atomic lattices. Since it is impossible to determine the chemical composition of

a mineral or the structure of its lattice without the use of complex and expensive apparatus, we have to fall back in practice on physical characteristics such as colour, hardness and so on as means of identification. Nevertheless it is important to remember that differences in these physical properties are really only expressions of fundamental chemical distinctions such as those outlined in Tables 2, 7 and 8.

Cleavage
The cleavage of a mineral is a direction of easy splitting, usually revealed when the mineral is crushed or dropped. Some minerals have no cleavage, some cleave in one direction, some in two or more directions, giving distinctive **cleavage fragments**; these variations are connected with the structure of the lattice. Cleavage in minerals must be distinguished from slaty cleavage in rocks, described on page 47.

Habit
The habit of a mineral is connected with the way in which it has grown. For example, minerals like asbestos (p. 54) which grow much more quickly in one direction than in others tend to form long thin crystals which are said to have an **acicular** (needle-like) or **fibrous** habit. Minerals like mica (p. 53) growing most quickly in two dimensions may have a **platy** or **flaky** habit. Minerals forming irregular or rounded sugary grains are said to have a **granular** habit. **Prismatic** minerals have an elongated form bounded by four or six parallel faces, as in the quartz shown in Figure 3.2. Minerals which have a **massive** habit show no distinctive shape and are not obviously crystalline.

Colour
Many minerals have a characteristic colour which helps in their recognition, but because small amounts of impurities may affect their colour it is difficult to lay down firm rules. In Tables 7 and 8 on pages 65 and 66 the most usual colour is given first. Variations in colour are much less apparent when the mineral is finely powdered. The colour of its powder is called the **streak** and can be determined by scratching the mineral with a file or knife, or by powdering a fragment with a hammer on a steel plate. The iron oxide haematite is an example of a mineral whose colour varies but whose streak is more constant. The unbroken mineral may be metallic grey or dark red, but the streak is always bright cherry red (p. 53).

Some minerals or varieties of minerals show very beautiful clear colours which give them a value as gemstones. **Ruby**, for example, is a deep red form of the aluminium oxide corundum. Even the common mineral quartz occasionally shows an attractive purple or pink colour (**amethyst** or **rose quartz** respectively) or a cloudy yellow, the latter characterising 'Cairngorms', found in parts of the Scottish Highlands.

Hardness
The hardness of a mineral depends on the tightness with which its atoms are packed together. Hardness may be estimated by rubbing the specimen over a fine file; a soft mineral powders easily and quietly while a hard mineral yields little powder and much noise. The noise and amount of powder can be compared with a standard set of minerals which provide a scale of hardness. Alternatively, the unknown mineral can be tested against the standards by finding out which standard minerals scratch or are scratched by it. A penknife will scratch minerals with hardness less than six, a fingernail minerals with hardness one to two. Diamond will scratch ordinary glass which has about the same hardness as quartz.

The standard minerals which provide *Mohs' scale of hardness* are:

Hardness	Mineral
1	talc
2	rock salt or gypsum
3	calcite
4	fluorspar
5	apatite
6	feldspar
7	quartz
8	topaz
9	corundum
10 (hardest)	diamond

Relative density
The relative density of a mineral is measured by its weight compared with that of an equal volume of water. Most silicate and carbonate minerals have relative densities of 2·5–3·5, but most metallic ore-minerals have densities greater then 4·0; galena (lead sulphide) has a relative density of over 7·0 and can easily be recognised by its weight (see Tables 7 and 8, pp. 65 and 66).

Lustre
The lustre of a mineral depends on the way in which it reflects light. Words used to describe lustre,

most of which make comparisons with other well known substances, include metallic (shown by many sulphide minerals), pearly, vitreous (like glass) and adamantine (like diamond).

Silicate minerals

With this information as a background, we can now consider the common silicate minerals. They are dealt with in the order in which they are listed in Table 2 (p. 20). Table 7 on page 65 is intended to summarise the details and also gives the essential facts about a few less important minerals mentioned in earlier pages. You should check the characters of mineral specimens against the table and try to decide which properties are of most use for diagnostic purposes.

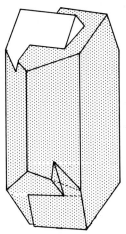

Figure 7.3 A twinned feldspar showing two feldspar crystals intergrown with each other.

Quartz (silicon dioxide, SiO_2). Quartz sometimes forms six-sided prisms capped by six-sided pyramids (Fig. 3.2), but when it occurs in rocks such as granite it more often forms shapeless grains. It is usually colourless (though coloured varieties do occur, as shown on p. 52) and often so nearly transparent that a broken chip looks like glass. Quartz has no cleavage and breaks with a conchoidal fracture (see p. 29). It has a low density, and is too hard to be scratched by a knife. Since oxygen and silicon are two of the most abundant elements of the Earth's crust, quartz is among the commonest of minerals. The lack of colour, clear glassy appearance, hardness of seven, threefold symmetry and conchoidal fracture distinguish it from other common minerals.

Feldspars. Feldspars are a family of silicates of aluminium containing calcium, sodium or potassium. The feldspars are the most abundant family of rock-forming minerals, the two chief species being **orthoclase** feldspar ($KAlSi_3O_8$) and sodium–calcium feldspar or **plagioclase** [$(CaNa)(AlSi)_4O_8$]. Feldspars usually form crystals with roughly the proportions of a brick and two almost rectangular cleavages. They are opaque and pale in colour: potassium feldspar is pink or white; plagioclase is usually white, yellowish, or grey. Both types sometimes show an iridescent play of colours including bright green or blue tinges. Many feldspars have the property of **twinning** – two or more portions of the crystal lattice grow up side by side in a definite geometrical pattern. A twinned orthoclase feldspar consisting of two individual crystals is shown in Figure 7.3. Twinned plagioclase feldspars often consist of a large number of individuals whose edges show up as fine striations on the surface of the mineral.

Mica. All micas contain aluminium, potassium and hydroxyl. White mica or **muscovite** has no iron or magnesium, black mica or **biotite** has both. Micas crystallise as flat crystals or flakes often six-sided in plan. A perfect cleavage runs parallel to the flakes and it is usually easy to peel off thin films from the surface of the mineral. Such cleavage flakes are flexible and elastic, that is, they spring back into shape after they have been bent, a property that makes micas easy to recognise. Both muscovite and biotite are translucent and sparkle in reflected light. Muscovite is almost colourless, biotite dark brown. Both are soft enough to be scratched by a knife, and both have a fairly low density. White mica has many industrial uses which depend on its flaky habit and on the fact that it is resistant to heat and is an electrical insulator.

Hornblende. Hornblende is a silicate containing calcium, magnesium, iron and aluminium with the hydroxyl (OH) ion. A member of the **amphibole** family of minerals, it crystallises as prisms showing two good cleavages, parallel to the prism axis, which meet at an angle of about 120° and serve to distinguish the mineral from augite (Fig. 7.4, see below). Hornblende is a dense opaque mineral, greenish-black and with faces which reflect light well. When it occurs in basic or intermediate plutonic rocks (cf. Fig. 5.9), it usually forms irregular grains. In hornblende schists (p. 48), the

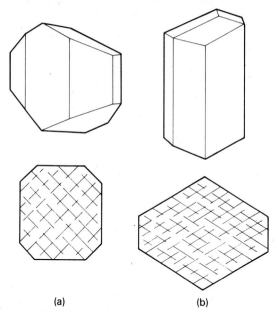

(a) (b)

Figure 7.4 (a) Augite, (b) hornblende. Above: crystals of common types; below: the two cleavages seen in cross section.

Figure 7.5 Garnet.

mineral often shows an acicular or needle-like habit. The tendency of hornblende to form elongated crystals is more pronounced in some other members of the amphibole family. It reaches its extreme in **asbestos**, one of the substances used commercially for insulation, in which individual crystals are thread-like.

Augite. Augite contains calcium, magnesium, iron and aluminium. It crystallises in squat prisms, rather square in cross section and often capped by sets of slanting faces (Fig. 7.4). Augite is usually black and opaque, with shiny faces which reflect light well. It is dense, fairly hard, and tends to break along two cleavage directions arranged almost at right angles parallel to the long axis of the prisms. The angle between the cleavages distinguishes augite from hornblende, as Figure 7.4 shows. Augite is a member of the important family of silicates known as the **pyroxenes**.

Olivine. Olivine is a silicate of iron and magnesium. It sometimes crystallises as short prisms capped by pyramids, but more often forms ovoid or rounded grains in basic or ultrabasic igneous rocks. The colour varies from yellow–green to dark green; the gem variety (peridot) is beautifully translucent. Olivine is dense, hard, and has a poor cleavage.

Garnet. Garnet contains two or more of the metals magnesium, iron, calcium, aluminium and chromium. It crystallises as many-faced crystals with three axes of four-fold symmetry. Garnets are hard, dense, and have no cleavage. Most garnets, including the gem varieties, are red, but some rarer types are brown or green. The red colour and the complex crystal form – like a ball with many facets (Fig. 7.5) – make garnet an easy mineral to recognise.

Non-silicate minerals

Although most of the common elements of the Earth's crust are contained in silicate minerals, many of the less common elements, including most of the metals used by man, form chemical compounds of different types which can be grouped as carbonates, oxides, chlorides or fluorides, sulphates and sulphides (Table 2, p. 20).

Carbonates

Calcium carbonate is the chief component of limestones and is the principal substance forming the shells of many animals. Two minerals, calcite and aragonite, have the formula $CaCO_3$ but differ in lattice-structure and crystal symmetry. A third mineral, dolomite, resembling calcite but containing magnesium in addition to calcium, is common

NON-SILICATE MINERALS 55

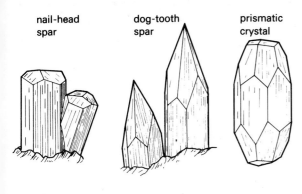

Common crystal forms of calcite

nail-head spar   dog-tooth spar   prismatic crystal

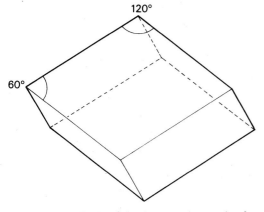

A cleavage fragment of calcite

The vertical axis of the fragment is an axis of three-fold symmetry.

Figure 7.6 Calcite: three common types of calcite crystal and a cleavage rhomb.

in carbonate rocks. Calcite is the only one of these minerals which need be described. It forms stumpy or slender six-sided prisms capped by pyramidal faces of various kinds (Fig. 7.6). The mineral is usually colourless and may be pearly, translucent or transparent. The three-fold symmetry gives a superficial resemblance to quartz, but calcite can be recognised, first by the fact that it is soft and easily scratched by a knife and secondly by the fact that it has a perfect cleavage and breaks into multitudes of little rhombs (Fig. 7.6). Calcium carbonate, as calcite crystals or in limestones, reacts with dilute acid, whereas dolomite does not. The reaction involves the production of carbon dioxide bubbles, and limestones therefore 'fizz' when a drop of dilute acid is placed on them.

Oxides

The important ore minerals of iron and tin are the oxides. Oxides and hydrated oxides of iron are very common both in sedimentary deposits and in veins or irregular masses deposited from magmatic solutions. Magnetite (Fe_3O_4), present in most basic igneous rocks, generally crystallises in blue-black eight-sided (octahedral) forms. It is perhaps the most strongly magnetic of natural minerals – it deflects the needle of a compass held near it and small grains are attracted to a magnet. The first compasses used in navigation were natural lodestones made of magnetite. Haematite (Fe_2O_3) very often forms nodular aggregates of small crystals, hence its common name 'kidney ore'. Fresh haematite is steely-grey and metallic-looking in mass, but its streak is bright red and it often has a reddish crust. Limonite is a brown earthy-looking iron-rich material which consists of iron oxides combined with water. It is formed by the alteration of other iron-minerals at the Earth's surface and gives a yellow–brown tinge to many weathered rocks, especially in desert regions.

Cassiterite, or tinstone (SnO_2), the world's main source of tin, is a tetragonal mineral, forming short brown or black prisms which are square in cross section and are capped by four-sided pyramids.

Chlorides and fluorides

Halite, sodium chloride (NaCl), is the common salt used in cooking, and in a variety of manufacturing processes. It is formed mainly by evaporation of sea water in desert climates and may be deposited either directly on the sea floor or between grains of sediments fringing the sea when these are permeated by saturated salt solutions. Halite is a soft cubic mineral which may often be scratched with a finger-nail. It has a perfect cubic cleavage, often growing in cubes which appear to have hollow sides. It is colourless and transparent and is easily identified by its taste and its solubility in water.

Fluorspar, or fluorite (CaF_2), often occurs with galena, sphalerite and quartz in mineral deposits of the Pennines and elsewhere. It is a source of hydrofluoric acid and is also used in making glass and enamels. Fluorspar forms translucent to transparent cubic crystals (Fig. 7.1) often of a beautiful shade of amethyst blue or purple but sometimes colourless, yellowish or green. An octahedral cleavage truncates the corners of the cubes. 'Blue John' is a massive fluorspar rock made up of colourless and blue or purple layers which occurs in Derbyshire, and is used for making ornamental vases, paperweights, etc.

Sulphates

Calcium sulphate forms several minerals of which the most important is gypsum. This mineral crystallises at low temperatures in sediments such as clays containing decaying organic matter. It is also deposited from sea water or ground waters made salty by evaporation in desert climates – in other words, it is an **evaporite** mineral (see p. 29). It usually forms flattish crystals which cleave easily and are often so soft that a finger-nail will scratch them. It is colourless, translucent and may have a pearly lustre. Satin spar is a fibrous gypsum consisting of thread-like crystals growing in silky parallel bunches. Alabaster is massive and is used for ornaments and statues. **Barytes** (barium sulphate, $BaSO_4$) is a gangue mineral of veins (p. 57), not unlike gypsum in appearance but easily recognised by its much greater relative density (gypsum 2·3, barytes, 4·5). Gypsum is used in the manufacture of plaster of Paris, plasterboard and other building materials, barytes to add bulk to paper and cloth and in drilling muds.

Sulphides

The sulphide minerals are of special importance because they include some of the principal *sources of copper, zinc, lead, nickel, arsenic and mercury.* They are found in various geological environments and appear to have been formed by more than one geological agency. Sulphide veins and segregations are deposited within the crust by fluids of magmatic origin which probably derive much of their metal content from magmas generated in the mantle. Sulphide bodies associated with sedimentary rocks at or near the surface are formed through the activity of bacteria, or where decaying organic matter provides a reducing (that is, oxygen-free) environment in which sulphur most easily combines with metals.

The metallic sulphides as a group are dense minerals which usually have a metallic lustre and a silvery, brassy, coppery or blackish colour. Some species tarnish on exposure to air and many emit a sulphurous smell when heated.

Pyrite or iron pyrites (iron sulphide, FeS_2), is one of the commonest of minerals. It is found in most sulphide vein deposits and forms nodules or replaces fossils in sediments where organic matter has decayed in the absence of oxygen. Pyrite has cubic symmetry and forms brassy-yellow cubes

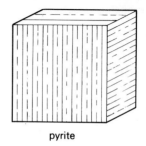

Figure 7.7 Iron pyrites or pyrite: a cubic crystal showing striated faces.

whose faces may be marked by fine striations (Fig. 7.7). It is sometimes mistaken for gold ('fool's gold'), but its cubic crystal form, hardness and brittleness make it easy to recognise. Chalcopyrite (copper pyrites, $FeCuS_2$) resembles pyrite in appearance but is coppery yellow and easily tarnished. The zinc sulphide, sphalerite (zinc blende, ZnS) is usually black or brown in colour. Galena (lead sulphide, PbS) is a very distinctive cubic mineral, lead-grey in colour, metallic in lustre and very dense (relative density is over 7). Nearly all of the lead used by man is derived from galena.

Native metals

Native (i.e. uncombined) gold, mixed with some silver, is occasionally found as **nuggets** or clastic grains in streams eroding gold-bearing rocks. Gold in such streams is concentrated by 'panning' – washing the stream sediment in a shallow pan over which water is allowed to flow. The high relative density of gold (19·3) ensures that it remains in the pan while lighter mineral grains are washed away.

Ore deposits

The minerals described above show that a number of the metallic elements of most importance to man are present in non-silicate minerals. The main sources of the world's gold, silver, copper, lead, zinc, tin and iron are provided by sulphides, oxides, and native metals. The ore minerals from which metals can be extracted are of special importance to the prospector and the mining geologist whose business it is to hunt for new supplies of ore. **Ore deposits** are simply rocks which contain a valuable metal (or metals) in such quantities that it can be extracted commercially. The **ore minerals** containing the desired element are usually mixed

ORE DEPOSITS 57

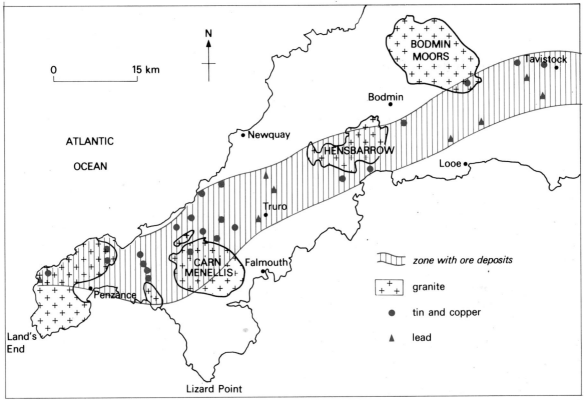

Figure 7.8 Ore deposits: the historic mining area of Cornwall, showing the distribution of deposits of tin and other metals in relation to granite intrusions. Most of the Cornish mines have now been worked out.

with unwanted minerals which miners call the **gangue minerals**, such as quartz, fluorspar or calcite, which have to be separated from the ore during the process of extraction.

The British Isles contain ore deposits of several kinds, some of which have been mined since long before the Norman Conquest in 1066. Deposits of tin, copper and lead occur in many veins and patches in Cornwall where they are concentrated around a number of granite stocks and in the roof of a granite batholith which is hidden below the land surface (Fig. 7.8). Lead deposits (containing galena) were formerly mined in the Pennines, the Lake District and the Mendips, and very rich deposits have recently been discovered in Ireland, some of which are located in limestone. Deposits of haematite were once mined in Cumbria; other types of iron ore have been mentioned on page 29.

Chapter 8

The Uses of Rocks and Minerals

For thousands of years, man has made use of rocks and minerals in ever-increasing variety. Historians call the earliest stage in the development of human communities in Europe the Old Stone Age (or palaeolithic age, to use the Greek words), because many tools and weapons characteristic of that period were made of flint or other rocks. If you look around your home or school (remembering that all metals are extracted from rocks and that most plastics and synthetic fabrics are made from petroleum), you will realise that our own way of life depends very largely on our ability to find and to process rocks and minerals which give us the raw materials we need. Some of the many rocks and minerals used today have been mentioned in Chapters 4 to 7. Table 6 brings together information about them in a grouping which emphasises the uses to which they are put, and the remainder of this chapter provides a short commentary on the table.

Fuels

Wood was the principal fuel of early man and is still widely used. Most of our industry, however, is powered by the **fossil fuels** (coal, oil and natural gas) which were formed many millions of years ago from buried plant and animal matter (pp. 30 and 33). Atomic or nuclear power from radioactive elements, of which uranium is the most important natural source, makes an increasing contribution in Britain and elsewhere. One of the great problems facing future generations is that of developing substitute power sources which can be used when supplies of the fossil fuels run out.

Building materials

Where suitable rocks are plentiful, stone-built houses roofed with slate (p. 47) characterise the landscape, especially in Scotland, Wales and Cornwall. In central, east and south-east England, the common building materials are bricks and tiles made from clay. Cement, made by heating mixtures of powdered limestone and mud or clay, is used to join bricks in brick-built houses. Buildings of the second half of the twentieth century, in all parts of the country, are made largely of concrete, which is made from cement together with crushed stone or gravel and is usually strengthened with a steel frame. Vast amounts of gravel and crushed stone (aggregate) are also needed for the foundations of roads, dams and other engineering works. The large quarries and pits which supply these materials can be seen all over the country. Bitumen derived from hydrocarbons (p. 33) is used in tarmac for surfacing roads.

Food production

Human communities crowded into large cities need reliable food supplies. Where the population is mainly rural, the needs are even greater because the average income in developing countries is usually low. Geologists help to maintain food production in two main ways. Mineral fertilisers such as phosphates and nitrates (Table 6) help to increase the productivity of the land. Still more important is the role of water supply. Fresh water is essential for both crops and livestock, and the provision of reliable supplies in many hot, dry countries has opened up huge new areas for farming. Some of the new water supplies come from aquifers located at depth (p. 32); much of the rest comes from rivers which have been diverted or dammed to feed permanent reservoirs.

Industrial raw materials

The metals used for tools and weapons in pre-

Table 6 Useful rocks and minerals.

PRODUCT	RAW MATERIAL	SOURCE
Fuels		
coal, peat	carbonaceous rock derived from forest or swamp debris	coal measures deposited on ancient coastal plains or deltas
oil and gas	fluids formed from decayed organic matter	obtained from porous reservoir rocks where trapped by impervious cap
atomic power	uranium from uranium minerals	veins or other segregations formed by waters circulating in crust
Building and road-making materials		
building stone	sandstone, limestone	sedimentary successions
	granite	plutonic intrusions
ornamental stone	granite, marble	metamorphosed limestones
roofing slate	slate	metamorphosed clayey sediments
tiles	clay	clayey sediments
bricks	clay	
cement	clay + powdered limestone	
concrete	cement + sand and gravel or crushed stone	sedimentary successions
'aggregate'	crushed igneous rocks, gravel, limestone	sedimentary and volcanic rocks
plaster and plasterboard	gypsum	mainly from evaporites
Fertilisers		
lime	limestone	sedimentary successions
potash, nitrates,	evaporites	
Industrial metals		
iron	iron oxides, carbonates	mainly from sedimentary successions
copper	sulphide deposits	concentrations mainly derived from magmatic fluids
tin	tin oxide	veins derived from magmatic sources
zinc	sulphide deposits	veins and concentrations deposited by circulating waters partly of magmatic origin
lead		
aluminium	bauxite, hydrated oxide	formed by weathering of clays in tropical climates
Gems and precious metals		
diamond	native carbon	igneous intrusions or placer deposits derived from them
ruby	aluminium oxide (Al_2O_3)	igneous intrusions or contact metamorphic rocks
sapphire		
gold	native element or gold tellurides	veins derived from magmatic sources, sedimentary deposits derived from erosion of such veins
silver		usually obtained as a by-product from deposits worked for lead, zinc, copper or gold

historic times were those that could be easily extracted from their ores – first bronze (which contains copper and tin) and later iron. Copper ores were mined in Cyprus more than three thousand years ago and tin was mined in Cornwall long before the Roman conquest of 55 BC. Pottery, made mainly from baked clay, has been in use in all but the most primitive societies for several thousand years. Modern technology makes use of a large number of metals, a few of which are listed in Table 6. In addition, many substitutes for metals, pottery and other natural products have been invented. Although these substitutes are man-made, most of them – polythene, nylon and many others – are derived from petroleum or from coal. Oilfields and coalfields are therefore the main suppliers not only of fuels but also of the petrochemical industry which is concerned with manufacturing these synthetic substances.

Ornaments

Lastly, we should remember how many objects of beauty come from rocks and minerals. The precious metals, gold and silver, and gemstones such as diamonds and rubies, have been among the treasures valued and sought for since prehistoric times. Many museums have displays of such treasures and if you live near London you may like to visit the exhibition of gemstones housed in the Geological Museum, Exhibition Road, London, SW7.

Chapter 9

The Earth as a Whole

The rocks of the Earth's crust, which have been dealt with in the main part of this book, are not distributed at random. We have seen, for example, that most andesites and most regionally metamorphosed rocks are formed in orogenic mobile belts (pp. 44 and 47). Furthermore, the orogenic belts that can be recognised in the Earth as it is today are very large structures which encircle the Earth (Fig. 9.3). These facts suggest that *the Earth's crust is affected by forces acting on a worldwide scale* which are responsible for variations in such factors as the temperature and mobility of the rocks at depth. In order to understand how those forces may work, we must end with a consideration of the Earth as a whole.

The crust, mantle and core

The Earth, one of the nine planets revolving around the sun, is made up of a number of layers of differing materials arranged one within the other. The outermost shell is gaseous and provides our **atmosphere**. The next shell, which does not completely envelop the Earth, is made of water and is represented by the seas, lakes and rivers. Inside this **hydrosphere** come two shells of solid rock – the **crust** and the **mantle** – and finally, a partially liquid **core** of iron and nickel. The materials which make the Earth are distributed between the shells in such a way that the average density of each shell increases inward from the atmosphere to the core (Fig. 9.1).

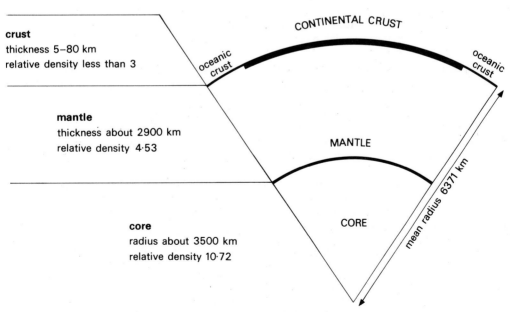

Figure 9.1 The core, mantle and crust of the Earth, a diagrammatic section.

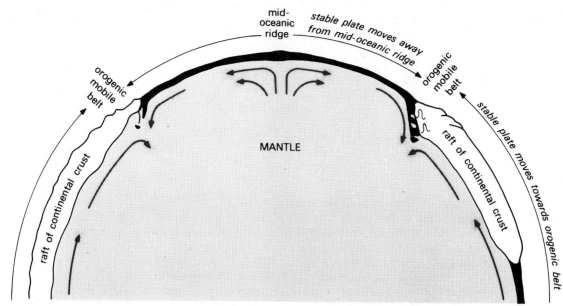

Figure 9.2 The effects of sea-floor spreading. The oceanic crust (black) grows wider as new igneous rocks are added to it. The rafts of continental crust are pushed aside by spreading of the oceanic crust, and begin to override the oceans at their leading edges. These movements are assisted by the slow circulation of plastic rock in the mantle. (Note: thickness of crust greatly exaggerated.)

The Earth can be measured and weighed, and such measurements show that it is more than five times as heavy as the same volume of water. The density of crustal rocks which can be seen near the surface, however, is seldom much more than three times that of water, and it follows that the mantle and core must be made of materials of high relative density as suggested in Figure 9.1.

The measurement of heat lost from the Earth's surface and the fact that high temperatures are encountered in deep mines show that *the Earth is hot inside*. The temperature reached near the base of the crust is usually several hundred degrees centigrade; that in the core perhaps well over 1600°C.

The shelled structure of the earth and the thicknesses of the crust and mantle have been established from studies of the behaviour of shock waves generated during earthquakes. Many kinds of vibrations are sent out from the focus of an earthquake, and after travelling through the Earth they can be picked up by sensitive instruments thousands of kilometres away. It has been found that vibrations of one type – the *S*-waves – are never received at observatories which lie almost opposite to the earthquake focus. Vibrations of this type cannot travel through liquids and it is therefore considered that part of the Earth's core, which blocks the *S*-waves, is in a liquid condition.

From the records received at observatories near to the focus of an earthquake, information has also been obtained about the outer shells of the Earth. *P*-waves – vibrations of another type – travel at constant rates so long as the material through which they move is uniform, but when they pass into rocks of different composition, they change both velocity and direction. The boundaries (**discontinuities**) which separate rocks of different kinds may reflect earthquake waves which strike them at certain angles, as light may be reflected by a mirror. The discontinuity that separates the mantle from the crust – which geologists call the **Moho** or **M-discontinuity** – has this effect. The depth of the Moho beneath the surface can be calculated by measuring the time taken by vibrations to travel down to the discontinuity and back again.

Putting together information supplied by the kind of observations mentioned above, it is possible to build up the picture of the Earth's structure illustrated in Fig. 9.1. The most important aspects of this picture from our point of view are those that concern the crust, and our next task is to go into these in a little more detail.

The continents and oceans

The first feature that catches the eye in any map of the Earth's surface is the contrast between the huge oceans on the one hand and the main land areas or continents on the other. Investigation of the structure of the crust has shown that this surface contrast is associated with very great differences at depth. The **continental crust**, which underlies both the continents and the shallow seas of the continental shelf at their margins, averages some 35 km in thickness, whereas the **oceanic crust** beneath the deep seas is usually only 5–10 km in thickness. This difference in thickness is balanced by a difference in average density: the relative density of rocks making the continental crust is about 2·7 to 2·9, whereas that of rocks making the oceanic crust is about 3·0 to 3·3. This arrangement allows the continents to 'float' with their surfaces well above sea level and their bases projecting far down into the mantle in much the same way as an iceberg floats in water (Fig. 9.2).

The contrast between the relative densities of the continental and oceanic crust results from differences between the most common rocks of each type of crust. The thick crust which forms the continents and continental shelves contains large amounts of granite (relative density 2·6–2·7) both as intrusive batholiths (p. 40) and as components of migmatites in regions of high-grade metamorphism (p. 49). On the other hand, granites are very rare in oceanic regions, where the principal rocks are basalts, formed by submarine eruptions (relative density of basalt 2·8–3·0). The predominance of basalts in the oceanic crust is illustrated by the fact that basalts are the main products of eruptions in Iceland, Hawaii and other volcanic regions which lie entirely within the ocean basins (Fig. 5.1). It thus appears that magmas formed at depth under oceanic regions must be mainly basic, whereas magmas formed under continental regions may be either acid or basic.

The mobile belts

Earthquakes and volcanic explosions show that, although the Earth feels solid and immovable beneath our feet, it is not at rest. Geological evidence shows that the crust has always been capable of slow movements: we saw in Chapter 2 that ancient sediments containing the fossilised remains of marine animals are found in the Himalayas where they have been raised thousands of metres above sea level. Earthquakes may displace a piece of the crust by a metre or so in a few minutes, and repeated displacements over the years may result in the shifting of parts of the crust for lateral distances of hundreds of kilometres.

Observation shows that at the present day both earth movements and vigorous volcanic activity take place mainly in a few rather narrow zones which cross both continents and oceans. The much larger stable areas between these mobile belts are seldom disturbed by earthquakes or volcanic eruptions. Following up the idea of mobility of the crust, we can recognise regions of three different kinds (Fig. 9.3):

(a) the oceanic ridge systems
(b) the orogenic belts } belts of crustal mobility.
(c) the stable plates

The oceanic ridge systems and orogenic belts are mobile belts of different kinds. Both are characterised by frequent earthquakes, by unusually high temperatures in the crust and by widespread volcanic activity. The **oceanic ridge systems** traverse the great oceans and are usually marked on the sea floor by broad submarine ridges cleft by central valleys; they occasionally approach the shore and invade the continents in narrow oceans such as the Red Sea. The **orogenic mobile belts** cross continental areas or follow the margins of the continents for long distances. They are characterised by rugged topography in which high mountain chains such as the Andes, volcanic island arcs (e.g. the Aleutian Islands) and deep trenches in the ocean floor are all represented (cf. Fig. 9.3 with Figs 2.5 and 5.1).

The **stable plates** which make up the rest of the Earth's crust may consist entirely of oceanic crust, entirely of continental crust or may be partly oceanic and partly continental. The plate that lies between the Mid-Atlantic ridge and the western coast of North America, for example, includes both continental (American) and oceanic portions. Most (though not all) of the stable plates are bounded at one side by an oceanic ridge and at the other by an orogenic belt. Earth scientists have discovered that each plate moves slowly as a whole away from the oceanic ridge at one margin and towards the orogenic zone at the other. We may think of the oceanic ridges as zones at which *the crust is being pulled apart*; no gap appears along these zones because molten magma from the mantle constantly rises to fill the opening spaces and the rise of this

64 THE EARTH AS A WHOLE

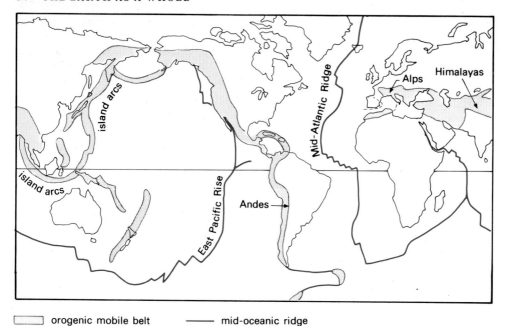

☐ orogenic mobile belt   ——— mid-oceanic ridge

Figure 9.3 The mobile belts and stable plates.

hot material builds up the sea floor into a ridge (Fig. 9.2).

The magma which feeds the oceanic ridges is basic in composition and forms basalts erupted on the sea floor, together with dolerites and gabbros intruded into the oceanic crust. As this magma solidifies, it forms a strip of new oceanic crust along the line marked by the oceanic ridge. In this way, each new phase of eruption makes the ocean basin a little wider, and the process of igneous activity at the oceanic ridges is therefore referred to as a process of **sea-floor spreading**. Accurate measurements show that the Atlantic Ocean is growing wider at a rate of a few centimetres a year. The continents on either side of the ocean – Europe and North America – are being pushed further apart as a result of this process of sea-floor spreading.

The orogenic mobile belts mark the lines at which **converging crustal plates** meet. The rocks of these belts are crumpled and broken by the collision, and excess crustal material is pushed down into the mantle where some of it begins to melt and stream up again towards the surface. We can see in these processes the explanations of the mobility of the crust in orogenic belts (Fig. 2.5) and of the concentration of volcanic activity along them (Fig. 5.1). The disturbance of the crust and the under-lying mantle brought about by collision causes temperatures to rise. The rocks in the deeper parts of the crust become folded and distorted by the stresses resulting from the collision and, as regional metamorphism begins, structures such as slaty cleavage and schistosity are developed (p. 47). At still deeper levels, portions of the ocean floor over-ridden by the continents in the collision zone are heated to temperatures at which melting begins. Magmas formed in this zone give rise to a variety of igneous rocks. Basalts, andesites and rhyolites may all be erupted to build up volcanic island arcs such as those that characterise the western side of the Pacific Ocean (Fig. 5.1). Granite magmas form large batholiths in the continental parts of the orogenic belt. Finally, the pressure resulting from the collision gradually forces the hot, distorted crust of the collision zone upwards to form a long, narrow mountain belt. The **mountain-building (=orogenic) phase** produces new highland regions which are attacked by the forces of erosion and which, in turn, provide the raw materials for a new phase of sedimentation. Thus, one change in the Earth leads to another in a never-ending sequence of rock-forming events. The history of the Earth can be established by following out these events as they are recorded in the rocks themselves.

Reference Tables

Table 7 Silicate minerals.

Mineral	Composition	Habit and crystal form	Cleavage	Colour	Hardness	Relative density
quartz	SiO_2	usually granular; crystals prismatic with pyramidal terminations	none	usually colourless, transparent, rarely pink, purple, yellow or smoky	7	2·65
FELDSPARS						
potash feldspar	$KAlSi_3O_8$	often granular; crystals usually tabular	two almost at right angles	white, pink or buff; opaque in hand specimen	about 6	2·56–2·58
albite ⎱ plagio-	$NaAlSi_3O_8$ ⎱ plagioclase series					2·6
anorthite ⎰ clase	$CaAl_2Si_2O_8$ ⎰ contains Na and Ca					2·76
MICAS						
muscovite	$KAl_2(AlSi_3)O_{10}.(OH)_2$	flaky; crystals platy, six-sided	one perfect, gives elastic flakes	white	2–2·5	2·76–3·0
biotite	$K(MgFe)_3(AlSi_3)O_{10}.(OH)_2$			brown	2·5–3	2·7–3·1
AMPHIBOLES						
hornblende	hydroxyl-bearing silicate of Ca, Al, Fe, Na	elongated or fibrous; crystals prismatic	two at 120°	dark green	5–6	3–3·5
PYROXENES						
augite	silicate of Ca, Al, Fe, Mg	granular; crystals stumpy prismatic	two at 90°	dark green to black	5–6	3·2–3·5
olivine	$(MgFe)_2SiO_4$	granular; crystals stumpy prismatic	two, poor	green	6–7	3·2–4·3
chlorite	hydroxyl-bearing silicate of Al, Fe, Mg	flaky	one perfect	green	2	2·6–2·9
clay-minerals	e.g. kaolinite $Al_4Si_4O_{10}(OH)_8$	flaky	one perfect	whitish		2·6
garnet	silicates of Ca, Mg, Fe, Al, Cr	granular; crystals complex, equidimensional	none	red, brown or green	6·5–7·5	3·5–4·3
andalusite	island structure; Al_2SiO_5	prismatic, square cross-section	parallel to length of crystals	red, pink or colourless	7·5	3·1–3·3

Table 8 Non-silicate minerals.

(a) Metallic ore-minerals.

Metal	Mineral	Composition	Colour	Streak	Cleavage	Hardness	Relative density	Common form, etc.
iron	magnetite	Fe_3O_4, iron oxide	iron-black	black	poor	6	5·18	octahedra, magnetic
	haematite	Fe_2O_3, iron oxide	steel-grey red or black	bright red	poor	6	4·9–5·3	tabular, kidney-shaped, earthy
	pyrite	FeS_2, iron sulphide	brassy-yellow	greenish black	—	6	4·8–5·1	striated cubes or massive
tin	cassiterite	SnO_2, tin oxide	black or brown	light grey or brown	—	6–7	6·8–7·1	massive, granular, water-worn grains
copper	chalcopyrite (copper pyrites)	$CuFeS_2$, copper iron sulphide	brassy-yellow, iridescent tarnish	greenish black	—	3·5–4	4·1–4·3	massive, softer than pyrite
	malachite	$CuCO_3.Cu(OH)_2$, copper carbonate	bright green often banded	pale green	—	3·5–4	3·9–4	massive, stalactitic
zinc	blende or sphalerite	ZnS, zinc sulphide	black, brown, rarely pale	white to brown	good	3·5–4	3·9–4·2	tetrahedra, massive
lead	galena	PbS, lead sulphide	lead-grey	lead-grey	cubic, perfect	2·5	7·4–7·6	cubes, massive or granular

(b) Other non-silicate minerals.

Mineral	Composition	Colour	Cleavage	Hardness	Relative density	Common form, etc.
gypsum	$CaSO_4.2H_2O$, hydrated calcium sulphate	colourless, often transparent, may be stained red or brown	one, good	2	2·3	tabular crystals, often twinned; often massive, without crystal form
rock salt, halite	$NaCl$, sodium chloride	colourless, often transparent, but may be opaque, stained brown	good, cubic	2–2·5	2·2	cubes: often massive, without crystal form; salty taste, dissolves easily in water
fluorspar	CaF_2, calcium fluoride	purple, colourless, yellow, blue or green, transparent to translucent	good, octahedral	4	3–3·25	cubes
calcite	$CaCO_3$, calcium carbonate	white or grey, occasionally red, yellow or other colours	good, gives rhombs	3	2·7	prismatic, effervesces with dilute HCl

Table 9 Sedimentary rocks.

	Rock	Constitution	Distinguishing features	Colour
DETRITAL ROCKS	*Coarse fragmental rocks* (>2 mm)			
	conglomerate	pebbly, mainly rock fragments with finer matrix	pebbles rounded	usually brown or red matrix
	breccia		pebbles angular	
	Arenaceous rocks ($2-\frac{1}{16}$ mm)			
	sandstone	sandy, made dominantly of quartz grains with varying amounts of feldspar and other minerals or rock chips	quartz dominant	buff, brown, grey or white
	arkose		quartz and feldspar	brown or red
	greywacke		grains angular, many kinds and sizes	grey
	Argillaceous rocks ($<\frac{1}{16}$ mm)			
	clay	made dominantly of clay minerals (rich in Al_2O_3)	very fine, moist, plastic	usually blue, grey or green, rarely red or brown
	shale		very fine, hard, platy	
	mudstone		very fine, massive	
CHEMICAL-ORGANIC ROCKS	*Carbonate rocks*			
	limestone	calcium carbonate	variable, often crystalline, sometimes many fossil fragments limestone reacts with dilute acid	usually white, cream or grey
	dolomite limestone	calcium magnesium carbonate		
	Ferruginous rocks			
	sedimentary iron ore	rich in iron minerals with chert or clay	structureless, nodular or striped, very high relative density	rusty coloured or black
	Evaporites			
	rock salt	sodium chloride	soft, very soluble, salt taste	light-coloured, gypsum often white
	gypsum	calcium sulphate	soft	
	coal	carbonaceous rock derived from plant matter	very fine, dull, with shiny jet-like layers; low relative density	black, dirty

For environments of deposition see Table 3.

Table 10 Igneous rocks.

Rock	Composition	Grain size and texture	Essential minerals	Other common minerals	Colour	Relative density	Occurrence	Geological environment
peridotite	ultrabasic, high MgO, CaO; low SiO_2, Na_2O, K_2O	coarse, granular	olivine, augite (or other pyroxenes)	hornblende	dark, sometimes greenish	high (3·3)	large intrusions, often associated with gabbro	intrusive bodies in orogenic belts; with gabbros
gabbro	basic, high MgO, FeO, CaO; low SiO_2, K_2O, Na_2O	coarse, granular	plagioclase, augite	olivine, hornblende	dark or speckled	high	large intrusions, laccoliths, ring-complexes	intrusions associated with basalts
dolerite		medium	plagioclase, augite	olivine	dark	high	dykes and sills	associated with basalts mid-oceanic ridges,
basalt		fine, often porphyritic	plagioclase, pyroxene	olivine	dark	high	lava flows and high-level dykes and sills	oceanic islands, stable continents, orogenic belts
diorite	intermediate	coarse, granular	plagioclase, augite or hornblende	biotite	greyish speckled	medium	large intrusions, often associated with granite	orogenic belts, often associated with granites
porphyrite		fine groundmass phenocrysts of feldspar	plagioclase, augite or hornblende		greyish	medium	dykes and sills	in association with andesites
andesite		fine, often porphyritic	plagioclase, augite or hornblende		dark or grey	medium	lava flows, often associated with pyroclastic rocks	orogenic mobile belts, especially island arcs
granite	acid, high SiO_2, Na_2O, K_2O; low MgO, CaO, FeO	coarse, granular	quartz, plagioclase or potassium feldspar mica	hornblende	light, buff or pink	low (2·6–2·7)	large intrusions, batholiths, stocks etc.	orogenic mobile belts, large intrusions and in migmatites
microgranite		medium			light, buff or pink	low	veins and small intrusions	associated with granites
porphyry		fine, phenocrysts of quartz and feldspar	quartz, feldspar		light, buff or pink	low	dykes and sills	associated with granites or rhyolites
obsidian		glassy	often no minerals large enough to identify, sometimes phenocrysts of quartz or feldspar		dark	low	lava flows	associated with rhyolites
pitchstone								
rhyolite		fine, often shows flow structure			light, buff		lava flows and plugs often associated with pyroclastic rocks	volcanic centres in orogenic belts or in basaltic provinces

Table 11 Metamorphic rocks.

Rock	Parent rock	Colour	Grain size	Fabric	Common minerals	Occurrence	
spotted rock	most parent rocks give these types on contact metamorphism, but rocks of clayey composition give best developed spotted rocks and hornfelses	often grey	very fine except in knots	knots representing incipient new minerals in fine groundmass	chlorite, mica or andalusite in knots	outer zones	contact aureoles, products of contact metamorphism
hornfels		grey or brown	fine to coarse	compact, no schistosity	in clayey rocks, andalusite, micas; in basic rocks, hornblende, pyroxenes	inner zones	
fault breccia	all types	often reddened	angular fragments in fine matrix	broken rock fragments	usually no new minerals, but quartz or calcite cement common		products of dislocation metamorphism
mylonite	all types	usually pale	very fine	platy, sometimes shows broken pieces of earlier minerals	usually no new minerals		
slate	rocks of clayey composition, tuffs	blue-grey, grey or green	very fine	splits into thin sheets parallel to slaty cleavage	chlorite, muscovite	low grades	deep parts of orogenic belts, products of regional metamorphism
schist	rocks of clayey composition, basic igneous rocks and others	grey, brown green or silvery	medium to coarse	splits into sheets or lenses parallel to schistosity	micas, garnet in clayey rocks, hornblende in basic rocks	medium grades	
gneiss	most types	usually pale, basic gneisses dark	coarse	streaky arrangement of minerals gives foliation	feldspars with biotite, garnet, hornblende, pyroxene	high grades	
migmatite	most types	granitic part pale, rest dark	coarse	mixture of granite and metamorphic rock	as in gneisses	high grades	
quartzite	sandy rocks	white or pale	fine to coarse	massive, may split parallel to bedding	quartz	in most environments of metamorphism	
marble	limestones	grey, white, pale green, buff	fine to coarse	massive	calcite (micas, amphiboles)	in most environments of metamorphism	

Index

Abyssal seas 23
Acicular habit 52
Acid rocks 41–2
Aeolian sediments 27–8
Agglomerate 38
Aggregate 58, 59
Alabaster 56
Aleutian Islands 34, 35, 63–4
Alkaline rocks 41
Aluminium deposits 59
Amazon River 30
Amethyst 52
Amphiboles 20, 53–4
Andalusite 47
Andes 44, 63, 64
Andesite 41, 43–4, 64
Animals 21, 22
Apatite 52
Aquifer 32–3
Aragonite 20, 54
Arenaceous rocks 22, 27–8, *see also* sandy rocks
Argillaceous rocks 22, 28, *see also* clayey rocks
Arkose 22, 23, 27
Artesian basin 32–3
Asbestos 54
Asphalt 33
Atlantic Ocean 64
Atmosphere 61
Atomic lattice 18, 50
Augite 20, 54
Aureole, *see* contact-aureole
Axes of symmetry 50–1

Bacteria 56
Bahamas 29
Barytes 20, 56
Basalt 41–4, 63–4
Basic rocks 41–4
Batholith 40, 57, 64
Bauxite 22, 59
Beach 10, 23, 25, 27
Bedding 14, 23–4, 45

Bedding plane 23
Biotite 42, 53
Bitumen 23, 33, 58
Black shale 23, 28
Blende, *see* sphalerite
Blue John 55–6
Bomb, volcanic 10
Boulder-clay 46
Boulders 21–2, 26–7
Breccia 22, 23, 26
Bricks 28, 58
Bronze Age 60
Building-stone 18, 29, 44, 58, 59

Cairngorm 52
Calcareous ooze 23
Calcite 28–9, 51, 52, 54, 57
Caldera 38
Canary Islands 10
Cap rock 33
Carbonaceous rocks 30
Carbonate rocks 28–9
Carbonates 20, 54–5
Cassiterite 55
Cement 29, 58
Cement, in sediments 25, 32
Central eruption 35–7, 40
Chalcopyrite 56
Chalk 28, 32
Chemical-organic sediments 22, 23, 28–31
Chert 22, 29
Chilled margin 41
Chlorides 20, 55
Chlorite 47
Clastic sediments 22, 26–8
Clay 22, 28
Clayey rocks 22, 28, *see also* argillaceous rocks
Clay minerals 20, 26, 28
Cleavage of minerals 50, 52, 53–4
Cleavage of rocks 47–8, 64
Coal 22, 23, 30–1, 58, 59
Coalfields 31, 60

Colour of minerals 52
Conchoidal fracture 29
Concrete 59
Concretion 29, 32
Conglomerate 22, 23, 26
Connate water 31
Contact-aureole 41, 46–7
Continental crust 44, 61–4
Continental shelf 23
Continents 12, 63
Copper deposits 57, 59, 60
Coral reef 24
Core 12, 61–2
Cornwall 47, 57
Corundum 52
Cotswolds 29
Crinoids 28
Cross bedding 24, 27, 28, 46
Crust 12, 19, 61–4
Crystal form 51
Crystals 18, 19, 50–1
Crystal symmetry 50–1
Cubic crystal 50–1, 56
Culbin Sands 10–11
Current bedding, *see* cross bedding
Cyclothem 31

Dartmoor 17
Deep-sea environments 23
Deltas 22, 23
Density of Earth 61–3
Density of minerals 50, 52
Denudation, *see* erosion
Derbyshire 56
Deserts 22, 23
Detrital sediments 22, 26–8
Diagenesis 25–6, 28, 29, 30
Diamond 19, 52, 59
Differentiation, sedimentary 21–2
Diorite 41–2, 44
Discontinuities 62
Dislocation-metamorphism 46–7
Dolerite 41–4
Dolomite 20, 54

INDEX

Dolomite rock 29
Dunes 10, 24, 28
Dyke 36, 37, 40

Earthquakes 11, 16, 17, 62, 63
Earth-movements 11, 16, 17, 63–4
Environments of deposition 22–3, 25
Environments of metamorphism 45–6
Erosion 15, 16, 21
Estuary 11, 23, 25
Evaporites 22, 23, 29, 55–6
Everest 16
Evolution 25
Extrusive rocks 34–40

False bedding, *see* cross bedding
Fault 11, 47
Fault-breccia 47
Feldspars 18, 19, 20, 52, 53
Felsic minerals 41–2
Ferromagnesian minerals 41–2
Ferruginous rocks 29, *see also* iron-rich rocks
Fertilisers 29, 58, 59
Fissure eruptions 35–7, 40
Flint 29, 58
Flood-plain 22, 23
Fluorspar, fluorite 50, 51, 52, 55–6
Folding 46, 48, 64
Foliation 48–9
Fossil fuels 58
Fossils 11, 16, 24–5, 28–9
Fuels 58, 59

Gabbro 41–2
Galena 52
Gangue 57
Garnet 20, 48, 54
Gas, *see* natural gas
Gas, volcanic 10, 35, 37–8
Gemstones 19, 52, 54, 59, 60
Giant's Causeway 44
Glaciation 23
Glaciers 22
Glass 38
Gneiss 47–9
Gold 56, 59
Grade, metamorphic 46–8
Graded bedding 27, 46
Grain-size 21–2, 40–1
Granite 18, 41, 44, 49, 51, 63, 64
Gravel 58
Gravity 21, 27
Great Barrier Reef 29
Greywacke 22, 23, 27
Groundwater 32–3
Gypsum 20, 29, 52, 56

Habit of minerals 52
Haematite 20, 28, 55, 57
Halite, *see* rock-salt
Hard water 32–3
Hardness 50, 52
Hawaii 34, 37, 44, 63
Highlands, Scottish 47
Himalayas 16, 17, 63
Hornblende 20, 53–4
Hornblende-schist 48
Hornfels 47
Hydrocarbons 28, 33
Hydrosphere 61
Hydroxyl 20

Ice-age 26
Ice sheet 26
Iceland 34, 63
Igneous rocks 15, 34–44
Ignimbrite 38
Interior drainage 23
Intermediate rocks 41–3
Intrusive rocks 15, 34, 40–1
Iron Age 60
Iron-rich rocks 22, 29, 57, 59

Japan 34, 35
Jurassic building stone 29
Juvenile waters 31

Laccolith 40
Lake District 57
Lakes 22, 23
Land's End 17, 57
Laterite 22
Lava 15, 35–40
Lava-plateau 35–6
Lead deposits 57, 59
Lime 59
Limestone 22, 23, 28–9
Limonite 55
Littoral environments 23
Local metamorphism 46–7
London basin 32–3
Lustre 53

Magma 15, 35, 37, 64
Magnetism 55
Magnetite 20, 55
Major elements 19
Mantle 12, 61–2
Marble 48
Mendips 57
Metamorphic grade 46–8
Metamorphic rocks 15, 45–9
Metamorphic zones 46–9
Metaquartzite 48
Mica 18, 19, 20, 53

Mica-schist 48
Microgranite 41, 44
Mid-Atlantic ridge 63, 64
Mid-oceanic ridge 42, 62–4
Migmatite 49, 63
Minerals 18, 20, 50–7
Mississippi River 11
Mobile belt 17, 23, 62–4
Mobility of crust 11
Moho 62
Mohs' scale 52
Mountain building, *see* orogeny
Mud, mudstone 13–14, 22, 28
Muscovite 42, 53
Mylonite 47

Nailhead spar 55
Native elements 56
Natural gas 28, 33, 58, 59
Neck 35–6, 40
Neritic environments 23
Nitrates 58, 59
Non-silicate minerals 20, 54–6, 66
Nuclear power 58

Obsidian 41, 44
Oceanic crust 61–4
Oceanic ridge-system, *see* mid-oceanic ridge
Oceans 12, 63
Oil 23, 28, 33, 58, 59
Oilfield 33, 60
Olivine 20, 54
Oölites 22, 28–9
Ore-deposits 56–7
Ore-minerals 52, 56
Organic matter 28
Ornamental stone 29, 59
Orogenic mobile belts 17, 44, 45, 46, 61, 62, 63–4
Orogeny 17, 64
Orthoclase (potassium feldspar) 42, 53
Oxides 20, 55

P-waves 62
Pacific Ocean 64
Peat 30, 59
Pebbly rocks 22, 26–7
Peléan eruption 37–8
Pennines 57
Peridotite 41–2
Permeability 32, 33
Persian Gulf 29
Petroleum 33
Phenocryst 40, 43
Phosphates 58, 59
Piedmont 23

Pillow-lava 38, 39
Pitchstone 41, 44
Plagioclase 42, 53
Plants 21, 22
Plaster 29, 56, 59
Plates, crustal 62–4
Plinian eruption 37–8
Plug 35–6, 40
Plutonic rocks 41–2
Pompeii 38
Pore-spaces 31, 33
Porphyrite 41, 44
Porphyritic rock 40, 43
Porphyroblast 48
Porphyry 41, 44
Potassium-feldspar 42, 53
Pressure in metamorphism 45
Prismatic habit 55
Pumice 38
Pyrite 20, 56, 57
Pyroclastic rocks 35, 37–8
Pyroxenes 20, 54

Quartz 18, 19, 20, 51, 52, 53, 57
Quartzite 23, 27, 48

Red clay 23
Red Sea 63
Reef limestone 22, 25
Regional metamorphism 46, 47–9
Residual sediments 22
Rhyolite 41, 44, 64
Ring complex 40
Ripple marks 9, 10, 27
Rivers 11, 22, 23, 26, 27
Rock-salt 20, 29, 52, 55
Rocks, defined 13, 18
Roofing materials 28, 48
Rose quartz 52
Ruby 19, 52, 59

Rudaceous rocks 22, 26–7

S-waves 62
Sand 10, 21
Sand-dunes 10, 28
Sandstone 23, 24, 26–8
Sandy rocks 10, 22, 26–8
Sapphire 19
Satin spar 56
Schist 47–8
Schistosity 48, 64
Scree 26
Sea-floor spreading 62–4
Seas 22–3
Seat-earth 30–1
Sedimentary rocks 14–15, 21–33
Shale 22, 23, 28
Shell sand 24
Shelly limestone 22
Shield volcano 37
Silica-percentage 41–2
Silicate minerals 20, 53–4, 65
Siliceous ooze 23
Siliceous rocks 29
Sill 36
Silver 59
Slate 47–8, 59
Slaty cleavage 48
Soils 21, 30
Sorting, *see* differentiation
Sphalerite 56
Spotted rocks 46
Spring 32
Stable crustal areas 17, 23
Stable crustal plates 62–4
Stock 40
Stone Age 29, 58
Stratigraphy 25
Streak 52
Stress in metamorphism 45, 47–8
Sulphates 20, 56
Sulphides 20, 56–7
Symmetry, *see* crystal symmetry

Talc 52
Temperature inside Earth 15, 35, 45, 49, 62
Textures 38, 42
Tidal flats 23, 25
Tiles 28, 58, 59
Till 26
Tillite 23, 27
Time, geological 12, 25
Tin deposits 57, 59
Topaz 52
Trace elements 19
Transport, sedimentary 21–2
Tuff 38
Turbidite 23, 27
Turbidity current 27
Twinning 53

Ultrabasic rocks 41, 43
Uranium 58, 59

Valleys 16
Vein 47
Vent 35–6, 39
Vesicle 38, 39
Vesuvius 34, 37–8
Viscosity 35, 37
Volcanic ash 38
Volcanic bomb 10
Volcanicity 15, 21, 34–8, 63–4
Volcanoes 10, 15, 34–8

Water-supply 31–3, 58
Water table 32–3
Weathering 15, 21–2, 26
Wind 10, 22, 27
Worm burrows 25

Xenolith 41

Zinc blende, *see* sphalerite
Zinc deposits 59
Zones of metamorphism 46–8